G. Resconi, L. C. Jain

Intelligent Agents

T0139987

Springer

Berlin
Heidelberg
New York
Hong Kong
London
Milano
Paris
Tokyo

Studies in Fuzziness and Soft Computing, Volume 155

Editor-in-chief
Prof. Janusz Kacprzyk
Systems Research Institute
Polish Academy of Sciences
ul. Newelska 6
01-447 Warsaw
Poland
E-mail: kacprzyk@ibspan.waw.pl

Further volumes of this series
can be found on our homepage:
springeronline.com

Vol. 138. S. Sirmakessis (Ed.)
Text Mining and its Applications, 2004
ISBN 3-540-20238-2

Vol. 139. M. Nikravesh, B. Azvine, I. Yager,
L.A. Zadeh (Eds.)
Enhancing the Power of the Internet, 2004
ISBN 3-540-20237-4

Vol. 140. A. Abraham, L.C. Jain,
B.J. van der Zwaag (Eds.)
Innovations in Intelligent Systems, 2004
ISBN 3-540-20265-X

Vol. 141. G.C. Onwubolu, B.V. Babu
*New Optimzation Techniques in
Engineering,* 2004
ISBN 3-540-20167-X

Vol. 142. M. Nikravesh, L.A. Zadeh,
V. Korotkikh (Eds.)
*Fuzzy Partial Differential Equations
and Relational Equations,* 2004
ISBN 3-540-20322-2

Vol. 143. L. Rutkowski
*New Soft Computing Techniques for System
Modelling, Pattern Classification and Image
Processing,* 2004
ISBN 3-540-20584-5

Vol. 144. Z. Sun, G.R. Finnie
Intelligent Techniques in E-Commerce, 2004
ISBN 3-540-20518-7

Vol. 145. J. Gil-Aluja
*Fuzzy Sets in the Management of
Uncertainty,* 2004
ISBN 3-540-20341-9

Vol. 146. J.A. Gámez, S. Moral, A. Salmerón
(Eds.)
Advances in Bayesian Networks, 2004
ISBN 3-540-20876-3

Vol. 147. K. Watanabe, M.M.A. Hashem
*New Algorithms and their Applications to
Evolutionary Robots,* 2004
ISBN 3-540-20901-8

Vol. 148. C. Martin-Vide, V. Mitrana,
G. Păun (Eds.)
Formal Languages and Applications, 2004
ISBN 3-540-20907-7

Vol. 149. J.J. Buckley
Fuzzy Statistics, 2004
ISBN 3-540-21084-9

Vol. 150. L. Bull (Ed.)
Applications of Learning Classifier Systems,
2004
ISBN 3-540-21109-8

Vol. 151. T. Kowalczyk, E. Pleszczyńska,
F. Ruland (Eds.)
*Grade Models and Methods for Data
Analysis,* 2004
ISBN 3-540-21120-9

Vol. 152. J. Rajapakse, L. Wang (Eds.)
*Neural Information Processing: Research
and Development,* 2004
ISBN 3-540-21123-3

Vol. 153. J. Fulcher, L.C. Jain (Eds.)
Applied Intelligent Systems, 2004
ISBN 3-540-21153-5

Vol. 154. B. Liu
Uncertainty Theory, 2004
ISBN 3-540-21333-3

Germano Resconi
Lakhmi C. Jain

Intelligent Agents

Theory and Applications

 Springer

Professor Germano Resconi
Department of Mathematics and Physics
Catholic University
Via Trieste 17
25100 Brescia
Italy
E-mail: resconi@numerica.it

Professor Lakhmi C. Jain
University of South Australia
Knowledge-Based Intelligent
Engineering Systems Centre
Mawson Lakes
5095 Adelaide
Australia
E-mail: l.jain@unisa.edu.au

ISBN 978-3-642-06031-1   e-ISBN 978-3-540-44401-5
ISSN 1434-9922

Springer-Verlag is a part of Springer Science+Business Media
springeronline.com

© Springer-Verlag Berlin Heidelberg 2010
Printed in Germany

Cover design: E. Kirchner, Springer-Verlag, Heidelberg
Printed on acid free paper   62/3020/M - 5 4 3 2 1 0

Preface

This research book presents the agent theory and adaptation of agents in different contexts. Agents of different orders of complexity must be autonomous in the rules used. The agent must have a brain by which it can discover rules contained within the data. Because rules are the instruments by which agents change the environment, any adaptation of the rules can be considered as an evolution of the agents. Because uncertainty is present in every context, we shall describe in this book how it is possible to introduce global uncertainty from the local world into the description of the rules.

This book contains ten chapters. Chapter 1 gives a general dscription of the evolutionary adaptation agent. Chapter 2 describes the actions and meta-actions of the agent at different orders. Chapter 3 presents in an abstract and formal way the actions at different orders. Chapter 4 connects systems and meta systems with the adaptive agent. Chapter 5 describes the brain of the agent by the morphogenetic neuron theory. Chapter 6 introduces the logic structure of the adaptive agent. Chapter 7 describes the feedback and hyper-feedback in the adaptive agent. Chapter 8 introduces the adaptation field into the modal logic space as logic instrument in the adaptive agent. Chapter 9 describes the action of the agent in the physical domain. Chapter 10 presents the practical application of agents in robots and evolutionary computing.

Our research is influenced by our so many visionary colleagues who have contributed directly and indirectly during the evolutionary phase of this book. We believe that the research ideas reported in this book will form a basis in the evolution of human-like thinking machines.

Thanks are due to Feng-Hsing Wang for his assistance during the preparation of the manuscript. We are grateful to Milesi Cornelia for proofreading of the manuscript.

Germano Resconi, Italy
Lakhmi Jain, Australia

Contents

Chapter 4.
Adaptive agents and complex systems

Chapter 5.
Adaptive agents and models of the brain functions

Chapter 6.
Logic actions of adaptive agents

Chapter 7.
The hierarchical structure of adaptive agents

Chapter 8.
The adaptive field in logical conceptual space

Chapter 9.
Adaptive agents in the physical domain

Chapter 10.
Practical applications of agents in robots and evolution population

Chapter One

Evolutionary Adaptive Agents

1.1 Introduction

The subject of this book is the agent theory or Adaptive Agents in different contexts. Agents of different orders of complexity must be autonomous in the rules used. The agent must have a brain by which it can discover the rules contained within the data. Because rules are the instruments by which agents change the environment, any adaptation of the rules can be considered as an evolution of the agents. As uncertainty is present in every context, we shall describe in this book how to introduce global uncertainty from the local world into the description of the rules. In conclusion adaptation of different orders, uncertainty and the rules obtained by the data, form a new entity that can simulate different natural phenomena.

1.2 Order of Contexts and Adaptation

One of the main challenges in modern science is presented by the study of complex systems. It is very difficult to give an exact definition of what a complex system is. As it is well known, there are many different ways to formalise the concept of complexity. Here we underline that, whatever definition of complexity we adopt, it can be always reduced to a measure of our difficulty in processing information relative to the system under study. This difficulty can be due to several different reasons:

(1) the information is unavailable as in uncertainty situations;
(2) the information is complete but available only in an implicit form, and, a very long time or a very great number of operations is required to make it explicit;
(3) the information is complete, available only in an implicit form, but, on principle, it cannot be transformed completely into an explicit form.

The first case is the one in which we have an incomplete knowledge of the system to be studied. Such a situation is common when dealing with economic, social, and cultural systems, and also when we study psychological or biological systems. The second case and the third one are common in the abstraction or global knowledge situation where we can describe only global concepts without having a deeper specification. In every case we can take a special domain of the complex system or context where it is possible to give rules and agents that link concepts. Every context is a local description of the complex system that we can control completely without uncertainties or conflicts. All agents that operate inside a context are provided with rules that are used to make structures and concepts. To describe concepts and contexts many tools are built as the semantic web, Resource Description Framework (RDF), conceptual maps and others.

Different orders of concepts are possible. Simple concepts as words, symbols and numbers are at the order zero. Examples of these concepts are numerical variables or names as Charles, Mary or objects such as plants, trees. Rules or predicates are concepts at the order one. Rules link concepts of order zero and they can be divided into two types. The concepts of type one are the domains of the rules and the concepts of type two are the concepts inside the ranges of the rules. Every concept of type one connected by a rule to a concept of type two is a semantic unity of the order one. Inside a context we can have many semantic unities connected one with the other to obtain a conceptual map by which an agent can navigate inside the context and move from one concept of order zero to another. A robot as an agent, with the map of a room, or conceptual map, can navigate inside the room, the context, and move from one position, concept at order zero, to another. An enzyme, as an agent, uses its rules to activate a chemical reaction which transforms molecules, concepts at order zero, into others. The whole process is inside a special context.

In a complex system we can have concepts of a higher order than the previous one. For example adaptation, evolution, drive plans are concepts at order two. The concept of order two transforms a set of rules inside a context, denoted context one, into another set of rules inside another context, denoted context two. The rule that relates context one with context two is a second order rule which is more complex than the rule of order one. Actually the agent in context two receives from context one, by means of rules of the second order, information of the rules in context one. This information is insufficient to rebuild the original rules from the point of view of the agent in the context two. The agent, with the information, uses its intelligence to build the original rules from its point of view. This process compensates the lack of information and eliminates every uncertainty on the original rules which now can be used in the context two. When the agent in context two receives information from its external context, it is in a conflict situation because its internal rules cannot fit in with the external information. Only when it uses the adaptation process or compensation process, it can adapt its rules to the new information. So in this case it eliminates every conflicting internal state. This necessity of using intelligence process of adaptation is the main difference between the rules of order two and rules of order one. The total compensation generates a symmetry situation between the context one and the context two. For example, when an animal, which is adapted to an environment, must adapt to another, it must change its physiological and cognitive system in such a way as to fit in with the new information received from the new external environment. The adaptation process of the animal or context two to the new environment or context one is a rule of the order two. The intelligence process inside the rule of the order two is the instrument by which the animal can survive in different environments. When the animal is adapted to its environment we can say that there is symmetry between the rules of order one in the environment and the rules of order one in the animal. Adaptation processes establish isomorphisms, homomorphisms or more complex connections between concepts of the order one in two or more contexts.

When one context is adapted to another context the conceptual maps in the two contexts are in general different. There exists a second order rule that transforms rules from one context to another. If the two contexts are not adapted one to the other, then the second context is not connected with the first. The second context is then separated from the first. To generate a suitable adaptation a set of agents is necessary. The Adaptive Agent will compare internal and external conceptual maps and generate a process by which all the different maps become equal. The second order conceptual map transforms rules from one context into the rules in another context. The second order conceptual map will assume that a process of compensation has been successful so that every context is connected with the others. In many cases it is impossible to adapt all the contexts at the same time. Usually only partial adaptation can be obtained. In this case we can have fuzzy adaptation for which "adaptation" has a certain degree of truth. The Adaptive Agents try, in a continuous way, to change the rules within the different contexts, in such a way as to enlarge the number of the contexts with the maximum degree of similarity. One example of this continuous adaptation is the historical evolution of physics where two fundamental contexts exist. One is the conceptual or mathematical context with the physical models. The other is the context of associated experiments. At the beginning the possible model of the experiments was very poor and the results of the theoretical computation were different from the experiment results. After the work of Galileo, Newton and many others, the models and the rules in the conceptual domain were changed or compensated in a dramatic way. Also the instruments to generate data were improved in such a way as to obtain good data. The improvement (adaptation) both of the theoretical model and of the experimental tests has obtained the result to join the theoretical context to the experimental context. We build a second order rule that implements the connection. In our historical example complete symmetry is obtained among all possible physical contexts in different positions of space and time. The theoretical model becomes valid in all the previous contexts. But at the moment many compensation problems are unable to be solved. The different forces of the universe which include gravity, electromagnetic, weak and strong forces are defined by different rules. At present there is not a symmetric rule that unifies all the different known forces.

The theoretical simulation is often in accord with the experiment. In the experimental context the external conceptual map and the empirical rules are the same. However we have cases where the conceptual rule and the empirical rule give two different results. Here it is necessary to change the adaptive agents, the physical theory and physical experiments in order to obtain identity between the abstract physical model and experimental tests. This is necessary in order to obtain symmetry between the concept and the physical experimental domain. The rule that joins the conceptual domain and the physical domain is of order two. The rules inside the conceptual domain or inside the physical domain are rules of order one.

Maps, rules, contexts at the order one, two, three, four or higher are possible. We know that the different orders are not separated one from the others. But one order includes all the smaller orders. For example there are contexts of the order two that include rules of the order two. These rules join contexts of order one. Inside the contexts of order one there are rules of the order one. Contexts of the order three include rules of the order three that connect contexts of the order two and so on. For contexts of the order three, four and so on, the process of compensation can be very complex. In fact a top down movement of compensation must be generated. Initially we compensate all the contexts of the order one so as to obtain global symmetry. We next compensate the contexts of order two. In this case when we change the rules of the order two we must restart from the order one and find again the lost symmetry. For any new higher order every adaptation destroys the adaptations at the lower orders which we must rebuild again. For example one animal, as an adaptive agent, can have micro adaptation and macro adaptation. For the micro adaptation the rules change for little changes of the environment and the animal needs to generate only small changes or adaptations in the different environments. When it moves to a completely different environment the previous micro adaptation must be rebuilt for the new environment. Every macro adaptation changes the micro adaptation. If the micro adaptation is at the order two, the macro adaptation is at order three. Every change or adaptation at the order three generates a new process of adaptation of the order two. Only when all adaptations for every order are obtained we can build the conceptual map and form the semantic unities at the higher order. In this case we obtain a super-symmetry for all possible contexts for every order.

To give a model of a complex system by means of contexts at different orders it can be useful to divide the complex system into simple parts or contexts. This operation of division is useful to compare the original system with the others. For example we have the complex system of the head. The head can be divided into many parts: eye, ear, brain and so on. Every part is a context of the order one. When we consider the eye context and the brain context for example, there is a situation of conflict between the eye and the brain because there exists a delay between the information in the eye and the information arriving at the brain. The brain receives an image which is not present in the retina. The brain forms an image after 100 ms which is not synchronous with the image in the eye. The image in the brain is not the same as the image in the eye and this is a conflicting situation between the two contexts. To avoid conflict the retina anticipates the future image by 100 ms and sends it to the brain. The computed image is transmitted to the brain and after 100 ms it arrives at the brain that can see the most probable image present in the retina after 100 ms. When the computation in the retina anticipates correctly the actual image the delay is eliminated and the brain can use the image at the same time as it is available in the retina.

When a complex system is represented by a set of variables in different spaces and with different dimensions, a conflict situation may arise among the different contexts. For example a gas can be represented by macro variables of pressure, volume and temperature. The macro context is structured by the well known macro rules that in thermodynamics are very simple. When we change the context of the gas and introduce a very high dimension giving the position and velocity of every molecule, the rules are still the traditional mechanical rules. In this case however we must introduce average values or special compensation rules of the order two to obtain symmetry between the micro and the macro context of the gas. Local contexts and global contexts can also be the source of asymmetry in the rules. In fact apparently simple recursion relations can hide unpredictable chaotic behaviours, as in the case of Feigenbaum's logistic map. A similar situation also characterises the world of mathematical logic and automata theory. The conflict in this case is between local and global representation of a system. In the local context the Feigenbaum's logicistic map is completely determined so that no uncertainty exists and locally the information is complete. But when a trajectory is generated we move from the local to the global context. Because at the global context we have unpredictable chaotic behaviours, the local and global contexts are not coherent. When it is possible, we can try to solve the conflict with a compensation process.

The various proofs which cannot be decided owing to Godel's theorems and the antinomies, show that, even if the information is complete, we cannot, on principle, reach a coherent definite context of the conclusions. Such a situation requires, when studying complex systems, an attitude which we could call as a context dependent logic valuation or possible world. We use this term to denote the fact that, when we deal with complex systems, we accept, as a starting point, that every model we utilise to describe them is, in principle, incomplete. It will be surely falsified by some future experimental findings that we obtain on the systems themselves.

When we accept that every model is incomplete or is context dependent, we break the symmetry and fuzzy uncertainty by the use of special fuzzy logic. Here we assume that compensation is impossible at the low level of the context of the order one. We then try to compensate at a higher order, second order or more and build higher order logic. In this way we avoid the illusion of the complete and universal compensation. Fuzzy set theory is the classical example in which we abandon the possibility of a complete compensation

1.3   Adaptive Agents at Different Orders

As we will see in chapter 3, adaptation is possible at different orders. At the first order we have a context with its concepts, rules, and agents. Every agent has a schedule to change the values of the variables in the context by the rules. At the first order no adaptation is possible. At the first order the semantic unity is made up of three elements.

[term1, rule, term2]

Term1 is the domain of the rule and term2 is the range of the rule. The semantic unity is the scheme by which we can represent the rule at the first order. In the natural language the unity at the first order is the subject as term1, the verb as the rule and the object as term2. The action of the first order agent is to transform one term of zero order into another.

At the second order the semantic unity is made by:

[context 1, second order rule, context 2]

Inside the context 2 we have two types of rules or conceptual maps. The first rule is the projection or transformation of the rule in the context 1 into the context 2 and is called external rule. The second rule is called the internal rule in the context 2. To avoid conflict or uncertainty the two rules must be equal. The second order agents or adaptive agents are designed to transform the two rules into one. When we fix the external rule and we change the internal rule, the adaptive agents compute the difference between the two rules. Using this difference the adaptive agents then adapt the internal rule so as to obtain a new rule for which the internal and external rules are equal. When the internal rule is fixed, the adaptive agents change the external rule. To change the external rule adaptive agents arise from the context 2. To obtain the internal identity, agents can change the original rule in the context 1 or the transformation of the rules from context 1 to context 2.

A simple example of the adaptive agent is the driver of a car. When the driver drives in a street without wind he is in the first context or context 1. When he drives in the street with wind he is in the second context or context 2. The rule in context 1 is to stay in the street. For context 1 the street is the rule. When he moves from one part of the street without wind to a part of the street with wind, or from context 1 to context 2, he cannot use the same rule. He must compensate his rule to stay in the street when there is wind. When he knows the rule for the second order which is a function of the velocity of the wind he can compute the new rule in the second context from the rule in the first context. He becomes an adaptive second order agent. To find the transformation from one context to the other it is necessary to have a lot of experience and samples so that the driver can automatically cope with the new situation having many different contexts dependent on different velocities of the wind.

1.4 Adaptation, Agents and Uncertainty

Given a context for the previous conditions, it is possible for the context to have two different rules, the internal rule and the external rule. When the two rules are equal, no conflict exists and uncertainty is eliminated. When we cannot equate the two rules, we must have two terms generated by the two rules. To study uncertainty, the rules in a context are represented using different points of view or worlds. In each world there is a set of rules that come from an external source or from an internal source. All the rules in the different worlds are conflicting rules that compete one with the other to represent the sole rule in the context. A context can be connected with many other contexts, so we can have many external rules that enter the context and only one internal rule. All these rules can be considered as the same rule but in different worlds. In this way a high uncertainty situation exists in a context. Inside the context symmetry is broken many times. Because symmetry establishes a bridge between the different contexts and the context that we study, the break in the symmetry generates uncertainty among all the contexts. In this situation we can use new forms of fuzzy logic to study the case.

To describe uncertainty we use the meta-theory based on modal logic. With this new approach, we define operations AND, OR and NOT between uncertainty structures that can be represented by fuzzy sets. The operations between the fuzzy truth sets become sensitive to the logic value true or false that agents, persons, sensors assign to the worlds. The operations are also sensitive to the difference of the worlds and time of synchronisation. It should be pointed out that when we use the logic operations AND, OR and NOT, we generally assume that the worlds are the same and change their truth-value at the same time. But it is known that there are cases where this synchronous situation and identity among worlds are not always valid. That is, the same proposition is transformed into one world at a certain moment and in another world at another moment different from the previous one.

In conclusion, the linguistic AND, OR, NOT operations become dependent on the particular truth value of a world, on the synchronisation and on the worlds assigned to the two propositions. Thus all of these possible changes in the structure of the worlds, in the modal logic, cause gradation of the linguistic operations as AND, OR, and NOT. An individual world (person, agent, sensor .) assigns to a sentence either a true or false value and uses the classical two value logic operations of AND, OR, NOT. That is the crisp true or false responses (assignments) of worlds to generate gradation of truth value. The uncertainty in a fuzzy set is represented with sets of worlds in a conflict situation. That is the same proposition may be true in one world and false in another. The linguistic operations change for different concrete situations (set of worlds). Whereas membership values of the fuzzy set of truth verifications associated with the set of worlds are in [0,1]. Furthermore, the combination of fuzzy membership values generates a Type II fuzzy set that is captured by FDCF and FCCF formulas. In a series of papers (Türksen 1986, 1994, 1995, 1999) a new approach to the representation of knowledge with canonical forms has been developed. It is shown that Fuzzy Disjunctive and Fuzzy Conjunctive Canonical Forms, FDCF and FCCF, respectively, determine the lower and upper bounds in knowledge representation. The interval-valued representation defined by FDCF and FCCF expressions are in part due to the representation of "linguistic terms" and in part to the combination of these terms with "linguistic operations", e.g.,

"AND", "OR", "IMPLICATION", etc., where both the "terms" and the operators admit information granulation. Furthermore, Türksen (1999) has pointed out that there are words that "describe" a property of an element with an "atomic sentence" of a natural language and there are words that "verify" the "truth" of a given "atomic sentence" either in the two-valued, {T, F}, paradigm or in the infinite (fuzzy)-valued, [T, F], paradigm. It is further pointed out that, when an "atomic sentence" is to be "verified" in the infinite-valued (fuzzy) paradigm, such verification statements become "atomic sentences" for further verification in two-valued paradigm in a hierarchy of verification (Resconi et all 1992). Sometimes, the elicitation of "fuzzy verification", i.e., "fuzzy truth" of descriptions may be obtained via the modal logic framework. Other times, the elicitation of "fuzzy truth" may be obtained with the pooled estimation method based on yes/no responses to an atomic sentence (Türksen, 1999).

We present a modal logic framework for the acquisition of "fuzzy truth", i.e., membership values of truth, within the framework of the Kripke modal logic. In this framework, it is shown how one computes the membership values of the FDNF (Fuzzy Disjunctive Normal Form) and FCNF (Fuzzy Conjunctive Normal Form) expressions based on the modal logic paradigm. In the world representation of the uncertainty inside the special context of the Adaptive Agent, we can generate a special type of logic by which we make logical computation on the rules whose logic value depends on the special world. Compensation will be expressed by transformations Γ of the true value for the same rule in different worlds. In fact the last goal of the compensation is to eliminate any uncertainty or asymmetry among all the worlds in the same context. In this case every context receives from the other contexts the same rule that is also equal to the internal rule. In all the worlds for the elimination of the conflicts by adaptation, the rules of the context have the same logic value. The rules are true for every point of view or worlds or are false. In the logical computation we also introduce the transformation Γ that can show the work of the Adaptive Agents or conversely can show the creation of a break in symmetry. When we assume that the external rules or the internal rules of a context are weighted in a different way, the weight is a field inside the set of worlds where there are the external and the internal rules. The weights are functions of the worlds and all the weights form one adaptive field.

Having defined the adaptive field we can assign to any possible world a degree of significance (weight). Extra information on the Adaptive Agent is the main source of significance. We can use the significance to express in a logic way the structure of the extra information. For example suppose we have three contexts that communicate one with the other and we know that one of the three is the most important. Given this context we have received two external rules and we have one internal rule. These are our three worlds. The world or point of view associated with the most important context has the maximum significance. In every context depending on the world's significance we can express in a logic way the structure of the extra information by means of which we know that one context is more important than the others. Our attention can then be focused on the most significant worlds or we can alternatively compare two or more different worlds one with another. In this way, a guide may be obtained to use information and to discover and measure the associated degree of uncertainty. The space of worlds is useful to divide information into important parts. Any sentence in these worlds may be evaluated for different orders of significance. An accessibility relationship exists between the most significant worlds and the least significant worlds. Modal logic models of information are possible with the adaptive field. Uncertainty is normally given without any logic but by means of fuzzy set theory. In this paper the adaptive field connects modal logic with fuzzy set theory to give a better description of uncertainty. The adaptive field is able to discover a new type of uncertainty. The mathematical description of the fields and of the adaptive field gives us a new possibility to obtain models of logic that depend on information.

1.5   The Brain of the Adaptive Agent

Agents act in a context by using rules. In order to generate rules or models, we use a neural network. The classical McCulloch and Pitts neural unity is widely used today in artificial neural networks and may be thought of as a non-linear filter. Classical neural networks (NN) are only capable of approximating a mapping between inputs and outputs in the form of "black box" and the underlying abstract relationships between inputs and outputs remain hidden. Motivated by the need in the study on neural architectures, for a more general concept than that of the *neural unity*, or *node*, originally introduced by McCulloch and Pitts, we introduce the concept of the *morphogenetic neural (MN) network*. We will show that differently from the classical NN. The MN network can encode abstract symbolic expressions that characterize the mapping between inputs and outputs, and thus show the internal structure hidden in the data. Because of the more general nature of the MN, the MN networks are capable of abstraction, data reduction and discovering, often implicit, relationships. With the proposed morphogenetic neural network it is possible to discover internal structure in data by finding (generally implicit) dependencies between variables and parameters in the model. Morphogenetic filter uses orthogonal and non-orthogonal n-dimensional basis functions; and the role of the scalar product in both the MN and in biological neural systems is discussed. With the scalar product of basis functions we can define an n-dimensional fundamental tensor that gives the properties of the space whose coordinates are the basis functions. The space of basis functions is the geometrical image of the data. When the fundamental tensor is a diagonal matrix the basis functions coordinates are an Euclidean Space. When the fundamental tensor is not a diagonal matrix we have a non Euclidean space and the distance between two points is not the classical Pythagorean quadratic expression.

The morphogenetic neuron uses two basis operations which are "Write" and "Read" operation, *i.e.*

(1) the operation (*"Write"*), starting from a suitable *reference space* and from the *constraints* to be satisfied generates the weights at the synapse,

(2) the computation operation (*"Read"*), starts from the weights to construct a suitable reply that satisfies the imposed constraints.

For the simplicity of the Morphogenetic Neuron we can use this tool to discover properties or rules within the data. When we send information between the contexts, we can use the Morphogenetic Neuron to obtain the rules from this information. The morphogenetic neuron or a network of morphogenetic neurons is the real Brain of the Adaptive Agent. The data in the brain of the Adaptive Agent are divided into two main parts. One part contains the most important data by which we generate the cognitive context or basis functions of the Agent. This context is an n–dimensional geometrical space that describes the important data. The other part of the data is all the other data that we can represent as a linear combination of the important data. All ordinary data are vectors that can be shown as the superposition of the basis vectors or basis functions. When we change the important data we move from one context to another and the brain joins the two contexts by a second order rule. The brain is used also to generate compensation or adaptation inside the second order rule

1.6 Conclusion

In this book we will present an integrated approach to the agent theory. Agents are considered at different orders. Simple agents act inside a context to build structures. Agents use an internal rule for this task. In many cases rules are difficult to be obtained and for this reason we propose a new model of neural network called a Morphogenetic Neuron which can find rules from data in a simpler way than the ordinary neural network. This is the brain of the Adaptive Agent. Agents at higher orders that work in contexts at the order two, three and greater orders are possible. We remark that the usual context is at the order one; the context at the order two joins contexts at the order one, the context at the order three joins contexts at the order two and so on. Because every evolution of an agent is an adaptation of one agent to the environment we can also examine autonomous evolutionary agents. With the meta-theory based upon modal logic we build agents that can manage uncertainty and that can make fuzzy operations such as AND, OR or NOT. Uncertainty can be computed by means of an Adaptive Agent.

Chapter Two

Adaptive Agents and Their Actions

2.1 First Order Agents

In this chapter we present the fundamental mathematical and cognitive scheme in which we locate the Adaptive Agent. We introduce the context definition at different orders, the transformations from one context to another and the adaptation processes by which we can establish a symmetric relation among the rules inside the different contexts. This chapter is not completely abstract but has the aim to introduce the concept of the Adaptive Agent. When we have only one context, adaptation is impossible. The definition of every context is made by variables and rules and the rules are fixed inside the context. The adaptation principle has its source within the change of the rules. Consequently because we have only one context and every change of the context is forbidden, it is not possible to have any adaptation process.

In conclusion when we have only one context we have only one family of rules and they cannot change one into the other. Every agent uses rules for its action so that every ordinary agent is of the order one and its adaptation freedom is equal to zero. Every automatic system or program for a computer is an agent of the first order. For a formal description of the rules and adaptation, it is useful to describe the elementary semantic unity. Every simple rule is the composition of a finite or infinite semantic unity represented symbolically in this way.

Table 2.1. Semantic Unity.

Statements	Resource	Property	Value
S_1	IN	P_1	OUT

Where S_1 is the input statement, IN symbolically represents the domain of the rule, P_1 the rule and OUT the range of the rule.

In Table 2.1 we denote the domain as the source of the data that the agent uses for the action. The rule is a property of the agent. So P_1 is the characteristic or the property of the agent. The range is the value of the action so OUT is denoted the value. An example of rule at the first order is the FIPA **contract net protocol** that we show in Fig. 2.1. [76]

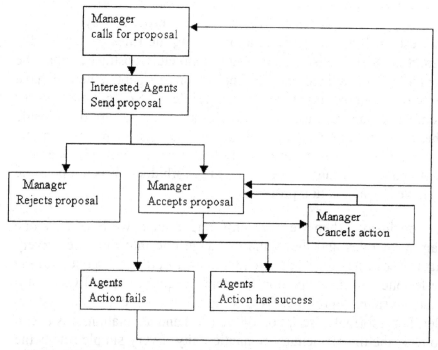

Figure 2.1. Communication rules between a manager and a group of agents.

In this contract net protocol the domain and the range of the rule are equal and are given by the elements:

IN = OUT = {"Manager calls for proposal", "Interested agents send proposal", "Manager rejects proposal", "Manager accepts proposal", "Manager cancels proposal", "Agent Action fails", "Agents Action has success"}.

In Figure.2.1 we show the relation that exists between the actions of two agents, the Manager and the interested agents. The manager, at the beginning, calls for a proposal for a given project. A group of interested agents sends proposals to the manager. He can reject or accept the proposals. When the manager accepts the proposal, the proposal can be cancelled by a cancelled action of the manager or because the agent fails in the realisation of the project. In any case the agent can send the manager a new proposal or can ask for a new acceptance action from the manager. The rule of communications between two agents in Figure 2.2 shows the context in which we operate. The context is only one and the Adaptive Agent cannot be successful because he must stay within the fixed rule of the context. One possible semantic unity is indicated in Table 2.2.

Table 2.2. Example of Semantic Unity for contract net protocol of Figure 2.1.

Statements	Resource	Property	Value
S_1	Manager calls for a proposal	P_1	Interested Agent Sends a proposal

2.2 Second Order Action for Adaptive Agent

When we divide the domain of the knowledge into parts or contexts, we have the possibility to express the knowledge by different types of rules. Inside every context we have fixed rules, but when we move from one context to another the rules can change.

Every agent must have knowledge of the context where it is to be located before using the rule for its action. With the context it will know the rule and the resulting action.

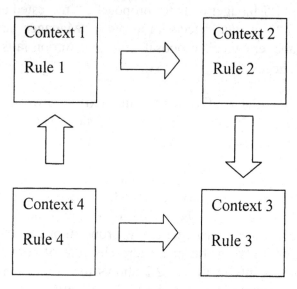

Figure 2.2. Network of contexts with its rules and communication channels.

Agents are divided into two orders. For the first order we have an agent which uses the rules, and for the second order we have agents that change the rules. We have presented the first order agent in chapter 2.1. The second order agent is presented in this chapter.

The semantic unity of agents of the second order or Adaptive Agent is given by the statements.

Table 2.3a. Semantic unity for second order agent.

Statements	Resource	Property	Value
S_1	IN_1	X_1	IN_2
S_2	IN_1	P_1	OUT_1
S_3	IN_2	P_2	OUT_2
S_4	OUT_1	X_2	OUT_2

Every semantic unity of the second order agent changes the rule X_1 in the first context into the rule X_2 in the second context. To simplify the notations, we replace the proposition:

"Second order agent" with the symbol "Agent^2 "

The rules P_1 and P_2 are the instruments by which the Agent^2 changes the rule X_1 into the rule X_2.

The semantic unity for Agent^2 can be graphically represented as follows.

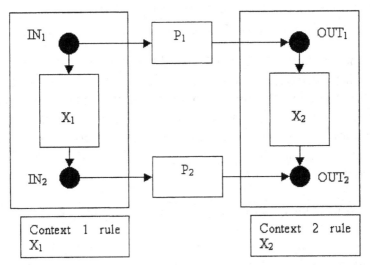

Figure 2.3. Description of the semantic unity for the second order agent that changes the rule X_1 in the context 1 into the rule X_2 in the context 2.

Figure 2.3 represents in a graphic way the statements shown in Table 2.3a. When we look at Table 2.3a, we note that S_1 and S_2 have the same source IN_1 in the context 1 and S_3 and S_4 have the same value OUT_2 in the context 2. This characteristic can be formally written as a relationship between the properties of the second order agent (X_1, P_1, P_2, X_2). The statements can be represented as follows:

$S_1 \rightarrow P_1 \ IN_1 = OUT_1$
$S_2 \rightarrow X_1 \ IN_1 = IN_2$
$S_3 \rightarrow P_2 \ IN_2 = OUT_2$
$S_4 \rightarrow X_2 \ OUT_1 = OUT_2$

We have the relationships between the properties:

$$P_2 \ IN_2 = X_2 \ OUT_1 \quad \text{or} \quad P_2 \ X_1 \ IN_1 = X_2 P_1 \ IN_1 \qquad (2.1)$$

The relationship (2.1) is the *internal coherence* for the Adaptive Second order agent or "Agent^2".

When the relationship (2.1) is false the adaptive agent is ambiguous. So it is incoherent and it does not exist one and only one relation X_2 in the context 2. In the context 2 we have two relations $X_{2,1}$ and $X_{2,2}$, where the $X_{2,1}$ is the transformation of the relation X_1, and relation $X_{2,2}$ is one of the possible relationships in the context 2. In the ambiguous situation the Agent^2 cannot use the relation X_2 because this is not unique.

A very simple case can be found in physics. When in the conceptual context or abstract context we create mathematical rules, and an Adaptive Agent then projects the theoretical models into the physical world, it must take account of the friction force. In many cases it is difficult to know the friction force or dissipative force, and so Agent^2 cannot always give a coherent model in the physical context. The rule X_2 in the physical domain must be divided into two rules $X_{2,2}$ and $X_{2,1}$. $X_{2,1}$ is the theoretical model X_1 and $X_{2,2}$ is the empirical or possible rule in the physical context when we have the friction force. We know that the two rules are different and this generates incoherence between theory and experimental tests.

2.2.1 Different Images of the Second Order Adaptive Agent

One of the possible schematic images of the semantic unity for Agent^2 is given in Figure 2.4.

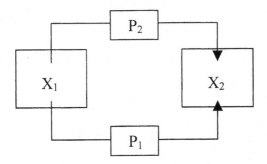

Figure 2.4. Schematic image of the semantic unity in Agent^2.

In a more mathematical way the same semantic unity can be represented in a slightly different way in Figure 2.5 where we have:

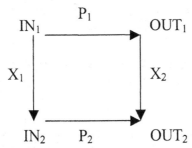

Figure 2.5. Mathematical image of the semantic unity in Agent^2.

Or, with a context image we can see the semantic unity in Agent^2 in Figure 2.6.

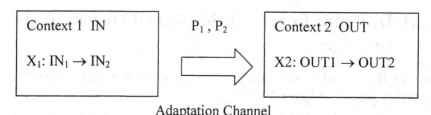

Adaptation Channel

Figure 2.6. Semantic unity in Agent^2 represented with context formalism.

We have shown X_1 and X_2 in Figure 2.6. For the statements in the Table 2.3a, the rule X_2 can be written in two different ways as shown in equation (2.2).

$$X_2 : OUT_1 \rightarrow OUT_2 \text{ or } X_2 : P_1 \, IN_1 \rightarrow P_2 \, IN_2 \tag{2.2}$$

When the Agent^2 is coherent the two forms give the same rule X_2, but when it is incoherent the two forms give two different rules.

A useful image of the Agent^2 and its semantic unity can be obtained by a special *linguistic* form created by G.J. Marshal, A. Behrooz [76]. The unity in the Table 2.3a and Figure 2.3 may be given in this way:

Activity: name of the statement of the second order
Environment 1 = Context 1
Environment 2 = Context 2
Object 1: variables IN_1 and IN_2 and methods X_1
Object 2: variables OUT_1 and OUT_2 and methods X_2
Aim of the adaptation channel: with P_1 and P_2 we build a channel to transform the object 1 into the object 2.

Example:

When we define the context and design the rule we have the second order action.

Activity: redesign
Environment 1 = Environment 2 = Context
Object 1: Existing object designed to a given specification
Object 2: required object with a different specification
Aim of the adaptation channel: Given the specification and design of Object 1 and the specification of the object 2, adapt the design of object 1 to give the design of object 2.

2.3   An Example of Coherent and Incoherent Agent[2]

When in context 1 we have the rule X_1 as shown in Figure 2.7, from one state we can pass and know the next one.

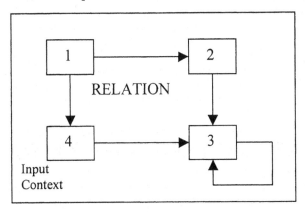

Figure 2.7. In context 1 or input context with the internal rule X_1 from an initial state we can compute the next state.

Within X_1 we have different paths. For example the path $1 \rightarrow 2 \rightarrow 3$, when we define $P = P_1 = P_2$ in the Table 2.4.

Table 2.4. Rule $P = P_1 = P_2$.

Statements	Resource	Property	Value
S_1	1	P	2
S_2	2	P	3
S_3	3	P	1
S_4	4	P	4

The projection of rule X_1 in the context 2 is rule X_2 that we show in Figure 2.8.

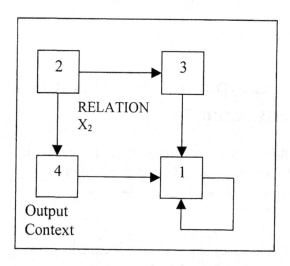

Figure 2.8. Rule in context 2 or Output Context.

When the values of the states inside the context 2 are (5, 6, 7, 8), the relation X_2 is incoherent with the context of the output or context 2.

2.4 Different Types of Actions for the Agent^2

2.4.1 First Type or Identical Transformation

In the first type for the action of Agent^2 relation X_1 in context 1 is the same in context 2. So $X_1 = X_2$ and thus the equation (2.1) can be written as follows.

$$P_2 X_1 IN_1 = X_1 P_1 IN_1 \tag{2.3}$$

To obtain the same rule X_1 in the second context the Agent^2 must balance the rules P_2 and P_1 so as to have the same rule X_1.

Example:

For example when rule X_1 is in the quadratic form:

$$X_1 IN_1 = a\, IN_1{}^2 + b\, IN_1 \tag{2.4}$$

Rearranging (2.3) we have:

$$P_2 [a\, IN_1{}^2 + b\, IN_1] = a\, (P_1\, IN_1)^2 + b\, (P_1\, IN_1) \tag{2.5}$$

Where $a\, (P_1\, IN_1)^2 + b\, (P_1\, IN_1) + c$ is the application of the same X_1 in the context 2. For (2.5) we must have:

$$P_2 IN_1{}^2 = (P_1\, IN_1)^2, \; P_2 IN_1 = P_1 IN_1$$

So we have $P_2 = P_1$ and $P_1 IN_1{}^2 = (P_1\, IN_1)^2$

In Figure 2.9 we show the identical transformation of the Agent^2.

Figure 2.9. Agent^2 creates the copy X_2 of X_1.

Example:

Physical example of the identity transformation: M. Jessel in 1973 wrote an important book [77] where he presented for the first time the global acoustic control by means of secondary sound sources. We consider that the work of M. Jessel was an initial approach to the Adaptive Second order agent. In the book, M. Jessel presented only the problem for the global control of the sound. However, one of the authors of the paper [14] G.Resconi extended the work of M. Jessel and named it General System Logical Theory. This can be considered the abstract and axiomatic treatment of the Adaptive Agent for the second or higher orders. In Chapter 9 completely dedicated to physics, we explain more fully M. Jessel's work and the application of the Adaptive Agent to physics.

In a physical space we know that every wave is generated by a source. The adaptive agent has two contexts: one is the physical medium where the sound propagates, the other is the context of the sources. We cannot confuse the source and the sound so we must use two different contexts.

Given that a physical medium such as air has a constant density ρ_0 in any point the infinitesimal mass of air has a velocity v and a pressure p. Here Q is the source of mass and F is the source of the momentum force in every point of the medium. We have in a simple case that the field of velocity and pressure of the air is regulated by the Euler conservative equation of the mass and of the momentum. Hence we have the system:

$$c^{-2} \partial_t p + \rho_0 \, \mathrm{div} \, v = Q \tag{2.6}$$

$$\rho_0 \, \partial_t v + \mathrm{grad} \, p = F \tag{2.7}$$

where (p, v) are the variables in the context of the physical field and (Q, F) are the variables in the context of the sources, c is the velocity of the sound.

When we transform the (2.6) in this way:

$$c^{-2} \partial_{tt} p + \rho_0 \, \partial_t [\mathrm{div} \, v] = \partial_t Q$$

and the (2.7) in this way:

$$- \rho_0 \, \mathrm{div} \, \partial_t v - \mathrm{div} \, (\mathrm{grad} \, p) = - \, \mathrm{div} \, F$$

We sum the two equations and have the famous D'Alambert equation.

$$c^{-2} \partial_{tt} p - \mathrm{div} \, (\, \mathrm{grad} \, p \,) = \partial_t Q - \mathrm{div} \, F$$

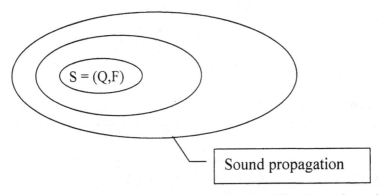

Figure 2.10. The propagation of the sound in the air by the surface with the same pressure and velocity. The distance between the surfaces corresponds to the second interval.

In Figure 2.11 the connection between the context of the physical field and the sources of the field is shown schematically.

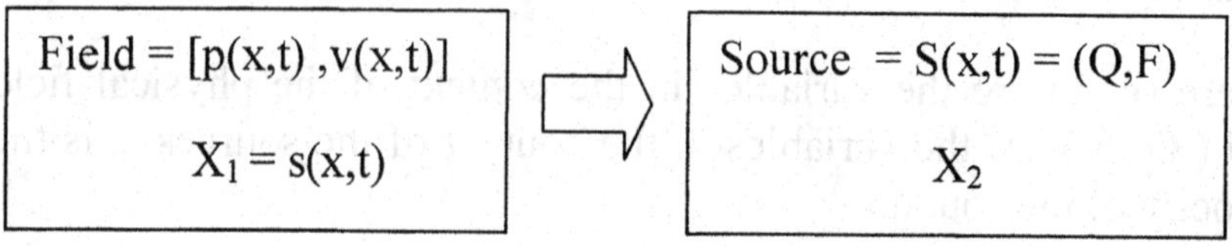

Field = [p(x,t) ,v(x,t)]		Source = S(x,t) = (Q,F)
$X_1 = s(x,t)$		X_2

Figure 2.11. The Agent[2] transforms the rule X_1 that changes the state of the air pressure and velocity into the rule X_2 that changes the source of the sound.

The second order agent that causes the transformation between the field and the sources can be represented by the statements as shown in Table 2.3b.

Table 2.3b. Semantic unity for second order agent.

Statements	Resource	Property	Value
ST_1	H = [p(x,t), v(x,t)]	s(x,t)	s(x,t) H
ST_2	H = [p(x,t), v(x,t)]	$c^{-2} \partial_t$ div grad $\rho_0 \partial_t$	$S_1(x,t) =$ (Q,F)
ST_3	s(x,t) H	$c^{-2} \partial_t$ div grad $\rho_0 \partial_t$	$S_2(x,t)$
ST_4	$S_1(x,t) = (Q,F)$	X_2	$S_2(x,t)$

When both the rules P_1 and P_2 in Figure 2.3 are equal to P, which is the system of (2.6) and (2.7), the Adaptive Agent uses P to transform:

$$p'(x,t) = s(x,t) p(x,t) \text{ and } v'(x,t) = s(x,t) v(x,t) \qquad (2.8)$$

that is X_1 in Figure 2.3 into the rule X_2 always in 2.3 which modifies the sources of the sound.

The first task is to consider the rules $X_2 = X_1 + X_{NL}$ where X_{NL} is the non linear rule that gives the difference with respect to the identical transformation for which $X_2 = X_1$.

The sources then have these transformations:

$$Q = s\,Q + Q_{NL}, \quad F = s\,F + F_{NL} \tag{2.9}$$

Substituting (2.8) in the system of equations (2.6), (2.7) and (2.9), we have this system:

$$c^{-2}\,p\,\partial_t s + \rho_0\,v\,\text{grad } s = Q_{NL} \tag{2.10}$$

$$\rho_0\,v\,\partial_t s + p\,\text{grad } s = F_{NL} \tag{2.11}$$

For the identical transformation $X_2 = X_1$ the non linear part is equal to zero, so (2.10) and (2.11) will be written as:

$$c^{-2}\,p\,\partial_t s + \rho_0\,v\,\text{grad } s = 0 \tag{2.12 a}$$

$$\rho_0\,v\,\partial_t s + p\,\text{grad } s = 0 \tag{2.13 a}$$

When we multiply the (2.12a) by $\partial_t s$ we obtain:

$$\partial_t s\,(c^{-2}\,p\,\partial_t s + \rho_0\,v\,\text{grad } s) = c^{-2}\,p\,\partial_t s\,\partial_t s + \rho_0\,v\,\partial_t s\,(\text{grad } s) = 0 \tag{2.12 b}$$

when we multiply the equation (2.13a) by $(-\,\text{grad } s)$ we obtain:

$$(-\,\text{grad } s)\,(\rho_0\,v\,\partial_t s + p\,\text{grad } s) = -\,\rho_0\,v\,\partial_t s\,(\text{grad } s) + p\,(\text{grad } s)$$
$$(\text{grad } s)\ = 0 \tag{2.13 b}$$

we sum up (2.12 b) to (2.13 b) we have the important equation:

$$p\,[c^{-2}\,(\partial_t s)^2 - (\text{grad } s)^2] = 0 \tag{2.14}$$

In a more simple case when p is different from zero we have the characteristic equation:

$$[c^{-2} (\partial_t s)^2 - (\text{grad } s)^2] = 0 \qquad (2.15)$$

Equation (2.15) is also called Eikonal equation which is the rule for free waves in a homogeneous medium where the direction of the wave velocity is perpendicular to the wave surface itself. When the Adaptive Agent begins with the rules X_1 for which (2.15) is true, we obtain the same transformation for the context 2. The identical transformation for which (2.14) is true describes the rule $s(x,t) = X_1$ inside the physical context (p, v) that divides the space (p, v) into surfaces that are normal to the velocity of the sound c. In conclusion when the adaptive agent transforms the rules X_1 in one context into the same rules in another context (source), the rules X_1 are the *natural rules* that describe the change of the context 1 (physical field) to *simulate the natural propagation* of the waves.

Another Physical Example:

Another example of identical transformation is presented. The equation of movement in mechanics is given by the well known Newton form:

$$F_i = M_0 \frac{d^2 x_i}{dt^2} = M_0 \, a_i \qquad (2.16)$$

Where F_i is the vector force, M_0 is the mass of the particle and a_i is the vector acceleration of the mass of the particle and x_i is the position of the particle. When we change the Newtonian reference where the velocity of the particle is much less than the velocity of the light c, into a reference where the velocity is near the velocity of the light, we know, by the Relativity Theory of Einstein, that the rule (2.16) no longer applies. How can we obtain new proper rules in the new context?

The equation (2.16) must be written as a rule for which for every passage from one context to another the rule is invariant; it must also be invariant if the terms of the rule change. The first idea is to change the vector position in this way:

$x_i = (x, y, z, ct)$ and the general force in this way $F = (F_1, F_2, F_3, W)$

where t is the time and W is the power [78]. When we choose the proper time τ for which the space time reference is located on the particle and the relative velocity is equal to zero, the Newtonian equation becomes:

$$M_0 \frac{d^2 x_i}{d\tau^2} = F_i \qquad\qquad (2.17)$$

Einstein assumed for his hypothesis that the relation (2.17) is invariant for every change of the space time or context with the change of velocity (Lorenz Transformation). In this case the Adaptive Agent uses the identity transformation for which:

$$X_1 = X_2 = \frac{d^2}{d\tau^2} \text{ is the same.}$$

Einstein was very lucky in this choice because all experiments show the validity of the identity transformation or symmetry principle.

2.4.2 Second Type or Transformation between Similar Rules

When the Adaptive Agent uses the rules $P_1 = P_2 = P$, the rule X_2 is not equal to the relation X_1 but is similar. Every context sends its similar image to the other. For the relations (2.1) we have:

$$P_1 X_1 = X_2 P_1 \tag{2.18}$$

With the statement image the action of the second order agent is given in the Table 2.5.

Table 2.5. Action of the Agent[2] which connects similar rules between contexts.

Statements	Resource	Property	Value
S_1	IN_1	X_1	IN_2
S_2	IN_1	P	OUT_1
S_3	IN_2	P	OUT_2
S_4	OUT_1	X_2	OUT_2

In the context image we show the rules within the different contexts.

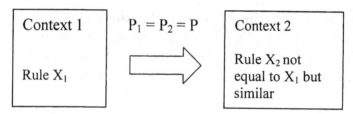

Figure 2.12. Context image of the second type that is the transformation between similar rules.

The fundamental question in the second type of action for the Agent[2] is to know the meaning of the word "similar" in the second type of transformation.

Definition: A rule X_2 is similar to a rule X_1 when there exists a transformation P that changes the domain and the co-domain of X_1 in such a way that the rule X_2 joins the transformed domain with the transformed co-domain. The rule X_2 is a projection of the rule X_1 by P.

Property:

The rule X_2 obtained by the second type of transformation is not always equal to the rule X_1.

In fact the rule X_2 can be written as follows:

$$X_2 : P \ IN_1 \rightarrow P \ IN_2 \ \text{ or } X_2 : P \ IN_1 \rightarrow P \ X_1 \ IN_1$$

and,

$$X_1 : IN_1 \rightarrow IN_2$$

given $P \ IN_1$ that belongs to the domain of X_1, the rule X_1 gives the result:

$X_1 \ P \ IN_1$ which can be different from $P \ X_1 \ IN$

which is obtained by X_2.

In conclusion when P and X_1 commute we come back to the identical transformation of the first type 2.4.1.

We have two different aspects of the second type of transformation. One aspect is related to the situation where P is one-to one and onto, the other aspect is related to the situation where P is only onto but is not one to one.

Graphic Example when P is one to one and onto:

To show the action of the type 2 Adaptive Agent for which we obtain similar rules, it is useful to represent the rule in X_1 as a graph of four elements. This graph gives the rule of the connection among the four elements. The graph is shown in Figure 2.13.

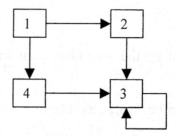

Figure 2.13. Relationships in the first context which give the rule by which we can move from one element to another. In fact with the rule X_1 we can move from 1 to 2 and 4 but we cannot move to 3.

The rule X_1 can also be given by the following table.

Table 2.6. Table representation of the rule X_1. The value 1 is the point where there is a relationship and the point where the value is zero the relationship is absent.

X_1	1	2	3	4
1	0	1	0	1
2	0	0	1	0
3	0	0	1	0
4	0	0	1	0

The relation in Figure 2.13 is a map that connects the four objects. When the active agent is used for $P_1 = P_2 = P$ the transformation is as shown.

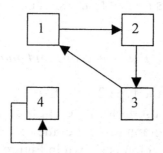

Figure 2.14. Rule P by which using the rule X_1 in the first context we obtain the rule X_2 which is similar to the rule X_1.

As in the case of the rule X_1, we can give a similar tabular representation for the rule P as shown in Figure 2.14.

Table 2.7. Table gives the representation of the rule P . In the position where the value is 1 there is the relationship. Where we have zero the relationship is absent.

P	1	2	3	4
1	0	1	0	0
2	0	0	1	0
3	1	0	0	0
4	0	0	0	1

Rule P of the second order Adaptive Agent enables the transformation of the rule X_1 into the rule X_2. Rule X_2 is shown in Figure 2.15.

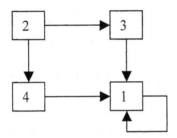

Figure 2.15. Rule X_2 obtained by the rule X_1 with the rule P.

Because the relation X_2 is defined in this way:

$X_2 : P\ IN_1 \rightarrow P\ IN_2$

So X_2 is,

$X_2 : P\ 1 \rightarrow P\ 2$
$X_2 : P\ 1 \rightarrow P\ 4$
$X_2 : P\ 2 \rightarrow P\ 3$
$X_2 : P\ 4 \rightarrow P\ 3$

$X_2 : P\,3 \rightarrow P\,3$

Or

$X_2 : 2 \rightarrow 3$
$X_2 : 2 \rightarrow 4$
$X_2 : 3 \rightarrow 1$
$X_2 : 4 \rightarrow 1$
$X_2 : 1 \rightarrow 1$

In tabular form we obtain,

Table 2.8. Representation of the rule X_2.

X_2	2	3	1	4
2	0	1	0	1
3	0	0	1	0
1	1	0	0	0
4	1	0	0	1

Because $P\,X_1$ is different from $X_1\,P$, the rule X_1 is different from the rule X_2.

In a graphic way the relationship X_2, when the objects 1,2,3,4 are ordered is given in Figure 2.16.

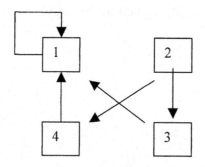

Figure 2.16. Rule X_2.

Remark:

When the rules in X_1 and in X_2 are similar there exists symmetry between the contexts. For the rules that are similar the structure of X_1 and X_2 is the same. We can associate one and only one path or process with every path or process obtained with the first rule in X_1. The structures of X_1 and X_2 are equal if X_1 and X_2 are different but similar.

Example:

In Figure 2.17 we show two paths, one in the rule X_1 the other in the rule X_2. It can be seen that the paths have the same structure but with different elements.

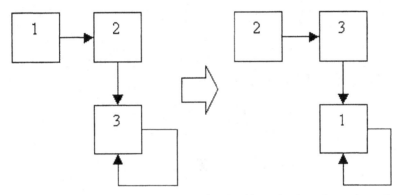

Figure 2.17. We have a path on the left using the rule X_1 and a path on the right using X_2, The rule X_1 is similar to the rule X_2 so with every path in X_1 we may associate a path with the same structure but with different values of X_2.

Graphic example when P is only "onto":

In the first context we have the rule whose graph form is shown in Figure 2.13. For P we use another P that is only "onto" but not "one to one". The rule P is:

$$P = (1 \rightarrow 2, 2 \rightarrow 2, 3 \rightarrow 1, 4 \rightarrow 1) \tag{2.19}$$

The transformed rule X_2 in the second context is similar to the rule X_1 but is not equal to X_1.

In Figure 2.18 we show the relation X_2 when P is "onto".

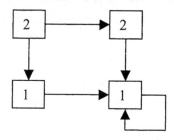

Figure 2.18. Rule X_2 is generated by the "onto" rule (2.19) from the rule X_1 in Figure 2.13.

It can be shown that if X_2 is similar to X_1 it is completely different from X_1. This is shown in Figure 2.19.

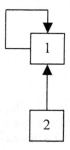

Figure 2.19. The same rule X_2 which is shown when we order the elements and joins equal elements is completely different from the rule X_1.

For similarity between X_1 and X_2: with a path in the rule we can associate an equal path where the elements are different and can also have repetitions. In Figure 2.20 we show the two paths in the two contexts.

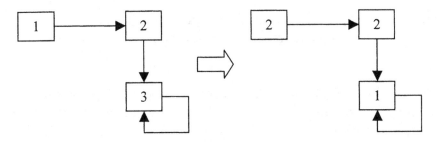

Figure 2.20. The paths in the context 1 on the left and the path in the context 2 on the right. The rule P is the onto rule in (2.19). We can see that with repetition and with a change of the objects the path on the left in an abstract sense is equal to the path on the right. Hence X_1 and X_2 are similar also when P is "onto". X_1 and X_2 have the same structure.

2.4.3 Third Type where Transformations between Rules are not Similar

When the Adaptive Agent uses the rules P_1 and P_2 where P_1 is different from P_2 the rule X_2 is not equal or similar to the relation X_1. The images sent from one context to the others are not equal or similar to the rules in the context. Even if P_1 is different from P_2, we can write the relation (2.1) in a general form:

$$P_2 X_1 = X_2 P_1 \qquad\qquad (2.20)$$

The statement image of the action of the second order agent is given in the following table:

Table 2.9. Action of the Agent^2 which connects rules X_1 and X_2 that are not similar.

Statements	Resource	Property	Value
S_1	IN_1	X_1	IN_2
S_2	IN_1	P_1	OUT_1
S_3	IN_2	P_2	OUT_2
S_4	OUT_1	X_2	OUT_2

The important property of this third type of transformation is that, it is possible to change (compensation) the rule X_2 in such a way that X_1 is similar to X_2 or $X_1 \approx X_2$.

We can also change the rule X_1 in such a way that the new rule X_1' is similar to the rule X_2. or $X_1' \approx X_2$. In fact equation (2.20) can be written in this way:

$$P_2 \, X_1 = X_2 \, G \, P_2 \quad \text{where} \quad P_1 = G \, P_2 \text{ and } G = P_1 \, P_2^{-1} \qquad (2.21)$$

The image of the action of the second order is given in Table 2.10.

Table 2.10. Action of the Agent[2] that connects similar rules between rule X_1 and rule X_2 G. Where X_2 G is the compensation rule of X_2 by means of the compensator term G.

Statements	Resource	Property	Value
S_1	IN_1	X_1	IN_2
S_2	IN_1	P_2	OUT_1
S_3	IN_2	P_2	OUT_2
S_4	OUT_1	X_2 G	OUT_2

For the second type of action in section 2.4.2, X_1 is similar to X_2 G or,

$$X_1 \approx X_2 \, G$$

Where G is the compensator term whose action changes the rule X_2. This is expressed in a graphic form in Figure 2.21.

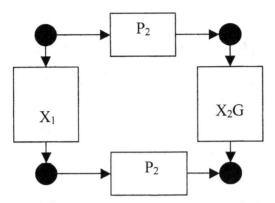

Figure 2.21. Graph showing the connection between X_1 and X_2 using the rule P_2 that joins the contexts and the compensator term G.

For equation (2.20) we can write this expression in the form of equation:

$$P_1\ H\ X_1 = X_2\ P_1 \quad \text{where } P_2 = H\ P_1 \quad \text{or} \quad H = P_2\ P_1^{-1} \qquad (2.22)$$

In Table 2.11 the action of the second order is expressed.

Table 2.11. Action of the Agent[2] that connects similar rules between rule X_1 and rule H X_1. Where H X_1 is the compensation rule of X_2 by the compensator term H.

Statements	Resource	Property	Value
S_1	IN_1	$H\,X_1$	IN_2
S_2	IN_1	P_2	OUT_1
S_3	IN_2	P_2	OUT_2
S_4	OUT_1	X_2	OUT_2

For the second type of action in 2.4.2, H X_1 is similar to X_2 or,

$$H\,X_1 \approx X_2$$

This can be expressed in a graphic way:

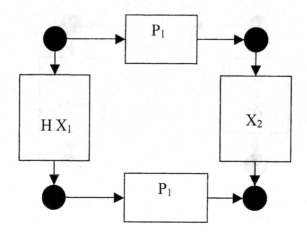

Figure 2.22. Graph that shows the connection between X_1 and X_2 using the rule P_1 which joins the contexts and the compensator term H.

Explanation of the compensator terms H and G by the use of internal and external rules.

When P_2 is different from P_1, the action or rule in the second context is expressed as an internal action and an external action.

When in the second context the agent receives a message it will form an internal elaboration and then an external elaboration. The internal elaboration is G and the external elaboration is X_2. The total computation X_2 G is similar (see (2.21)) to the rule X_1 but the individual computations of G or X_2 are not similar to X_1. In conclusion the similarity with X_1 is realised by the internal computation G that in this way compensates for the non-similarity of X_2 with X_1. Figure 2.23 shows the connection between G and X_2.

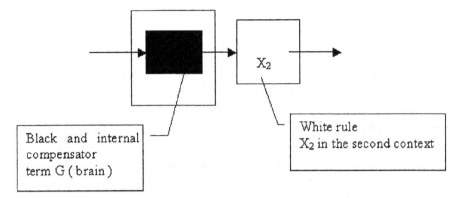

Figure 2.23. Represents the internal compensator term G (brain) and the main rule X_2 (body) in the second context. The composition rule of brain and body is X_2G that is similar to the rule X_1 in the first context.

All the previous ideas of internal and external rule may be repeated with the compensator term H (brain) as the internal compensator term, and X_1 as the external rule. In this case we have that $H X_1$ is similar to the rule X_2 (see (2.22)).

Figure 2.24 shows the connection between H and X_1.

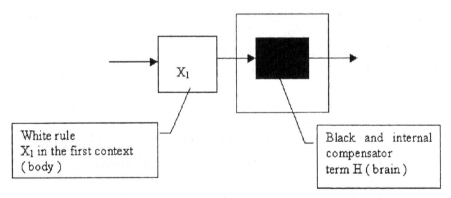

Figure 2.24. The internal compensator term H (brain) and the main rule X_1 (body) in the first context. The composition rule brain and body is $H X_1$ and is similar to the rule X_2 in the second context.

A Biological Example:

In biology a situation of non-similarity or conflict exists between the context of the eye and its rules or images X_1 and the context of the brain with its rules or images X_2. There exists a delay from the information in the eye and the information which arrives at the brain. The brain receives an image after a delay of 100 ms when it is not present in the retina. The brain elaborates an image after 100 ms delay and is not synchronous with that in the eye. The image in the brain is a virtual image. To avoid the loss of similarity between the image in the eye and the final image in the brain, the retina with the actual current image computes the likely image that will arrive at the retina after the 100 ms delay. The forecast program in the retina which simulates the image 100 ms later is called the compensator term or "H Brain". The computed image is received by the brain as $H X_1$.

When the rule H is a good model, the image on the retina after 100 ms is equal or similar to the computed image, that is after 100 ms the brain receives an image that is the same or similar to the actual image on the retina at that time. So thanks to the compensator program H in the context of the brain the image X_2 is similar to the image HX_1. The rule H is inside the retina but X_1 is the external image that arrives on the retina.

Graphic example:

Given the rule X_1 which is shown in Figure in 2.12 and where $IN_1 = IN_2 = \{1, 2, 3, 4\}$ we can write the following relationship.

$$X_1 = \{ 1 \rightarrow 2, 1 \rightarrow 4, 2 \rightarrow 3, 4 \rightarrow 3, 3 \rightarrow 3 \}$$

When we define the rules P_1 and P_2 of the Adaptive Agent in this way.

$P_1 = \{1 \rightarrow 2, 2 \rightarrow 3, 3 \rightarrow 4, 4 \rightarrow 1\}$

$P_2 = \{1 \rightarrow 3, 3 \rightarrow 2, 2 \rightarrow 4, 4 \rightarrow 1\}$

Because,

$X_2 : P_1 \text{ IN}_1 \rightarrow P_2 \text{ IN}_2$ we have,

$X_2 = \{P_1\ 1 \rightarrow P_2\ 2, P_1\ 1 \rightarrow P_2\ 4, P_1\ 2 \rightarrow P_2\ 3, P_1\ 4 \rightarrow P_2\ 3, P_1\ 3 \rightarrow P_2\ 3\ \}$

However for the rules P_1 and P_2 we have,

$P_1\ 1 = 2, P_1\ 2 = 3,\ P_1\ 3 = 4, P_1\ 4 = 1$ and,

$P_2\ 1 = 3, P_2\ 2 = 4,\ P_2\ 3 = 2\ , P_2\ 4 = 1$ so we have the result:

$X_2 = \{2 \rightarrow 4, 2 \rightarrow 1,\ 3 \rightarrow 2,\ 1 \rightarrow 2,\ 4 \rightarrow 2\ \}$

To compare X_1 with X_2 is useful to consider the graph of X_2 in Figure 2.25.

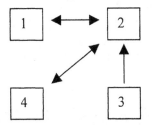

Figure 2.25. Picture of the rule X_2 in the second context. It can be seen that because P_1 is different from P_2 the rule X_2 is not only different from X_1 but it is also not similar. We can see that in X_2 there are different paths that do not exit in X_1.

In fact we have the path:$1 \rightarrow\ 2 \rightarrow\ 1$ with two steps that are not present in the rule X_1.

At this moment we build the compensator term G for (2.21) that can calculate X_2 G in the following way.

$G = P_1 P_2^{-1}$, because we know P_2, we can calculate the inverse of P_2^{-1}. This is possible because P_2 is a one to one and an onto relationship. We then have:

$P_2^{-1} = \{1 \rightarrow 4, 4 \rightarrow 2, 2 \rightarrow 3, 3 \rightarrow 1\}$ and

$G = \{1 \rightarrow 1, 4 \rightarrow 3, 2 \rightarrow 4, 3 \rightarrow 2\}$

Also

$X_2 G = \{1 \rightarrow 2, 4 \rightarrow 2, 2 \rightarrow 2, 3 \rightarrow 1, 3 \rightarrow 4 \}$

This can be obtained from X_2 by P_2 in the following way:

$X_2 G = P_2 IN_1 \rightarrow P_2 IN_2 = P_2 X_1 IN_1$

We also have,

$X_2 G = \{P_2 1 \rightarrow P_2 2, P_2 1 \rightarrow P_2 4, P_2 2 \rightarrow P_2 3, P_2 4 \rightarrow P_2 3, P_2 3 \rightarrow P_2 3 \}$

And because $P_2 1 = 3$, $P_2 2 = 4$, $P_2 3 = 2$, $P_2 4 = 1$ which gives the result:

$X_2 G = \{3 \rightarrow 4, 3 \rightarrow 1, 4 \rightarrow 2, 1 \rightarrow 2, 2 \rightarrow 2 \}$

That is equal to the computation by using G. By using G we can compensate X_2 in such a way as to obtain a rule similar to X_1. Using the Figure 2.26 we can show better the similarity between X_1 and X_2 G.

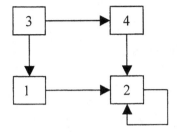

Figure 2.26. The rule $X_2\,G$ in the second context. The rule X_2 is different from X_1 but is similar because we can establish a one to one correspondence between the paths of X_1 and the paths of X_2. The rule G compensates the non similarity of rules X_1 and X_2.

In Figure 2.27 we show the internal rule G and external rule X_2 in the composition $X_2\,G$. We have:

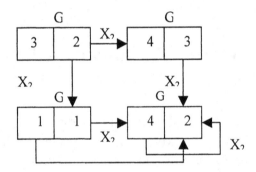

Figure 2.27. Diagram of the rule $X_2\,G$ in the second context. The operator G is located internally in the boxes, X_2 is external.

Remark:

Given two rules X_1 and X_2 which are not equal and not similar in two different contexts, when there exists a rule P_1 "one to one" or "onto" and another rule P_2 "one to one" and "onto", it is possible to compute a compensator term G for which $G\,X_2$ is similar to X_1. When P_1 is "one to one" and "onto" and P_2 is only "onto", it is possible to compute the compensator term H.

Example:

When we know the two rules

$$X_1 = \{1 \rightarrow 2, 1 \rightarrow 4, 2 \rightarrow 3, 4 \rightarrow 3, 3 \rightarrow 3 \}$$

$$X_2 = \{2 \rightarrow 4, 2 \rightarrow 1, 3 \rightarrow 2, 1 \rightarrow 2, 4 \rightarrow 2 \}$$

And

$$P_1 = \{1 \rightarrow 2, 2 \rightarrow 3, 3 \rightarrow 4, 4 \rightarrow 1 \}$$

$$P_2 = \{1 \rightarrow 3, 3 \rightarrow 2, 2 \rightarrow 4, 4 \rightarrow 1 \}$$

We can compute G and X_2 G which is similar to X_1.

2.5 Third Order Actions in Adaptive Agents

When we divide the domain of the knowledge into cognitive parts, it is possible to divide the knowledge into rules of the first order, rules of the second order and rules of the third order.

Because we associate the action of one agent with every rule, we can also have action at the first order, at the second order and at the third order.

Definition:

Every third order rule changes second order rules into second order further rules.

When we divide the knowledge into different contexts, we have contexts of the first order which include rules or methods of the first order, contexts of the second order which include rules of the second order and contexts of the third order which include rules of the third order.

Groups of contexts of the first order with rules of the first order change with a third order relation as we can see in Figure 2.28.

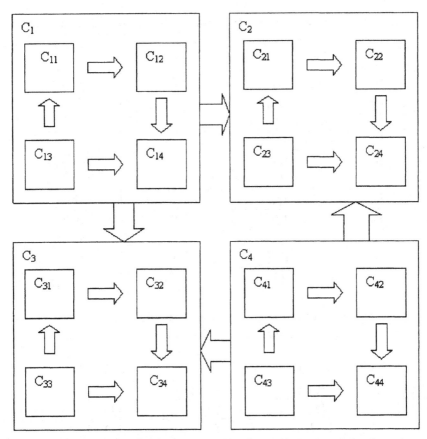

Figure 2.28. The internal rules that transform the contexts $C_{h,k}$ into the contexts $C_{i,j}$ are rules of the second order. These rules change one rule of the first order inside the context $C_{h,k}$ into another rule of the first order inside $C_{i,j}$. The large external rules that change the meta contexts C_1, C_2, C_3, C_4 one into the others are rules or actions at the third order where rules or actions at the second order in C_1 are changed into rules or actions of the second order in C_2 and so on.

Agents are divided into three orders of agents. At the first order we have agents that use the rules, at the second order we have agents that change the rules and at the third order we have agents that control the change of the rules. In the Table 2.12 we show the third order action of the Third order agent or $Agent^3$.

Table 2.12. Action of the $Agent^3$ or semantic unity of the third order agent.

Statements	Resource	Property	Value
S_1	IN_{11}	X_1	IN_{21}
S_2	IN_{11}	P_1	OUT_{11}
S_3	IN_{11}	A_1	IN_{12}
S_4	IN_{21}	P_2	OUT_{21}
S_5	IN_{21}	A_3	IN_{22}
S_6	OUT_{11}	X_2	OUT_{21}
S_7	OUT_{11}	A_2	OUT_{12}
S_8	IN_{12}	X_3	IN_{22}
S_9	IN_{12}	P_3	OUT_{12}
S_{10}	IN_{22}	P_4	OUT_{22}
S_{11}	OUT_{21}	A_4	OUT_{22}
S_{12}	OUT_{12}	X_4	OUT_{22}

Every semantic unity of the third order agent changes the second order rule $X_{1,2}$ given by the Table 2.13

Table 2.13. Semantic unity for the second order relation $X_{1,2}$.

Statements	Resource	Property	Value
S_1	IN_{11}	X_1	IN_{21}
S_2	IN_{11}	P_1	OUT_{11}
S_3	IN_{21}	P_2	OUT_{21}
S_4	OUT_{11}	X_2	OUT_{21}

into the second order relation $X_{3,4}$ as shown in Table 2.14.

Table 2.14. Semantic unity for the second order relation $X_{3,4}$.

Statements	Resource	Property	Value
S_1	IN_{12}	X_3	IN_{22}
S_2	IN_{12}	P_3	OUT_{11}
S_3	IN_{22}	P_4	OUT_{22}
S_4	OUT_{12}	X_4	OUT_{22}

To simplify the notations, the proposition "Third order agent" is substituted by the symbol "Agent^{3}". The rules A_1, A_2, A_3, A_4 are the instruments by which the Agent^3 changes the rule $X_{1,2}$ into the rule $X_{3,4}$. The semantic unity for Agent^3 can be represented graphically in this way.

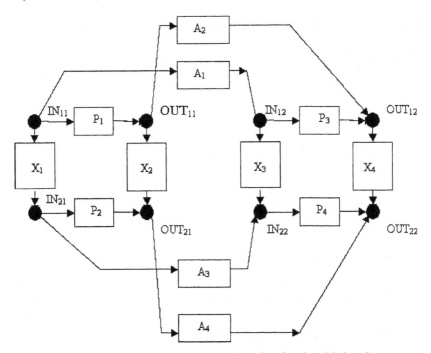

Figure 2.29. Description of the semantic unity for the third order agent that changes the rule $X_{1,2}$ or $X_1 \rightarrow X_2$ inside the context of the second order 1 into the rule $X_{3,4}$ or $X_3 \rightarrow X_4$ inside the context 2 of the second order.

Figure 2.29 represents in a graphic way the statements that we show in the Table 2.12. When we look at the statements in the Table 2.12, we note that the statements, S_7 and S_9 have the same value OUT_{12}, S_4 and S_6 have the same value OUT_{21}, S_5 and S_8 have the same value IN_{22}.

$S_7 \rightarrow A_2$ $OUT_{11} = OUT_{12}$
$S_9 \rightarrow P_3$ $IN_{12} = OUT_{12}$
$S_4 \rightarrow P_2$ $IN_{21} = OUT_{21}$
$S_6 \rightarrow X_2$ $OUT_{11} = OUT_{21}$
$S_5 \rightarrow A_3$ $IN_{21} = IN_{22}$
$S_8 \rightarrow X_3$ $IN_{12} = IN_{22}$

For the previous statements we have that

P_3 $IN_{12} = P_3$ $IN_{12} = OUT_{12}$, P_2 $IN_{21} = X_2$ $OUT_{11} = OUT_{21}$, A_3 IN_{21} $= X_3$ $IN_{12} = IN_{22}$

For the Table 2.12 we have the three equations:

A_2 $P_1 = P_3$ $A_1 = OUT_{12}$, P_2 $X_1 = X_2$ $P_1 = OUT_{21}$, A_3 $X_1 = X_3$ $A_1 =$
IN_{22} (2.23)

We also have that the statements S_{10} . S_{11}, S_{12} have the same value OUT_{22} in the context 2 of the second order. We then have:

$S_{10} \rightarrow P_4$ $IN_{22} = OUT_{22}$
$S_{11} \rightarrow A_4$ $OUT_{21} = OUT_{22}$
$S_{12} \rightarrow X_4$ $OUT_{12} = IN_{22}$

For the previous statements we have:

P_4 $IN_{22} = A_4$ $OUT_{21} = X_4$ OUT_{12}

For the equation (2.23) we have:

$$P_4 A_3 X_1 = A_4 P_2 X_1 = X_4 A_2 P_1 \qquad\qquad (2.24)$$

The relations (2.23) are the *internal coherence* for the three Adaptive Agents or "Agent^2 " included in the Adaptive Agent Action of the order three "Agent^3 ". The three agents Agent^2 always change IN_{11} as initial object. The relation (2.24) is the *internal coherence* for Adaptive Agent of the order three or "Agent^3". When one of the equations (2.23) or (2.24) of the adaptive agent is not true the transformation (action) of the Agent^3 is fuzzy.

With the context image we have that the action of the Agent^3 is located in the structure of contexts. We have shown this in Figure 2.29.

In Figure 2.30 we nest sets of contexts one within the other. One context is the source of the relations that are transformed when we move from one context to the other. The change of the original source of relations can act in three different ways which we denote as T_1, T_2 and T_3 in Figure 2.30. These are actions of the second order agent for which the equations are (2.23). We also have other two changes of rules B_1 and B_2, which converge on the same context and so one must be coherent with the other. The coherent equations for B_1 and B_2 are the equations (2.24).

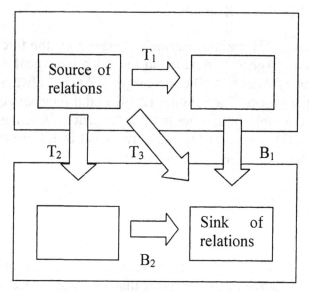

Figure 2.30. Here we show the actions of the agents at the third order. The arrows are actions that change the rules from one context to another. The arrows T_1, T_2, T_3 are actions of agents of the second order and these are the relations (2.23). The convergence of the two actions B_1 and B_2 on the same contexts is an action in which there are the two relations (2.24). These together result in an action of the Third order agent.

When we denote T_1, T_2, B_1, B_2 as statements that change the rules and not the objects, the Table 2.12 can be written in a more compact way as shown in Table 2.15.

Table 2.15. The Semantic unity in compact form of the action of Agent[3].

Statements	Resource	Property	Value
T_1	(IN_{11}, IN_{21})	(P_1, P_2)	(OUT_{11}, OUT_{21})
T_2	(IN_{11}, IN_{21})	(A_1, A_3)	(IN_{12}, IN_{22})
B_1	(OUT_{11}, OUT_{21})	(A_2, A_4)	(OUT_{12}. OUT_{22})
B_2	(IN_{12}, IN_{22})	(P_3, P_4)	(OUT_{12}. OUT_{22})

We note that the structure of (2.15) has the same structure as the structure in the Table 2.3 a. The relation that connects a third order adaptive agent with a second order agent is a homoeomorphism. The structure of the second order agent is contained in the third order agent. This may be seen when we substitute the objects with the transformations or actions of the first order. The Table 2.15 may be represented graphically as shown in Figure 2.31.

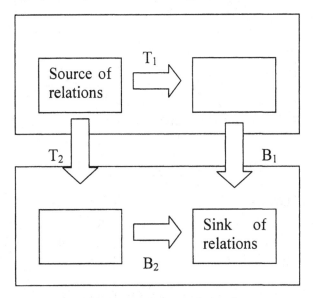

Figure 2.31. Actions of the agents of the third order. The arrows are actions that change the rules from one context to another. The arrows T_1, T_2, B_1 and B_2 are the actions of agents of the second order or of statements in the Table 2.15. The relation between second order agents and the third order relation is a homeomorphism.

In Figure 2.31 we show the actions of the agents of the third order. The arrows are actions that change the rules from one context to another. The arrows T_1, T_2, B_1 and B_2 are the actions of agents of the second order or of statements in the Table 2.15. The relation between second order agents and the third order relation is a homeomorphism.

In a more mathematical way the action of the third order agent can be represented by the cube with oriented arrows as shown in Figure 2.32.

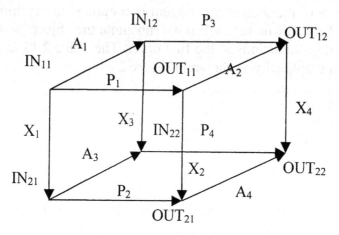

Figure 2.32. Abstract image of the action of the third order agent.

The action changes the second order action (IN_{11}, IN_{21}) → (OUT_{11}, OUT_{21}) into the second order action (IN_{12}, IN_{22}) → (OUT_{12}, OUT_{22}) by the rules (A_1, A_2, A_3, A_4). The rules inside the contexts in Figure 2.31 can be written in this way:

$X_1 : IN_{11} \rightarrow IN_{21}$ $X_2 : OUT_{11} \rightarrow OUT_{21}$ or $X_2 : P_1 IN_{11} \rightarrow P_2 IN_{21}$

$X_3 : IN_{12} \rightarrow IN_{22}$ $X_3 : A_1 IN_{11} \rightarrow A_3 IN_{21}$

$X_4 : OUT_{12} \rightarrow OUT_{22}$ $X_4 : P_3 IN_{12} \rightarrow P_4 IN_{22}$

$X_4 : P_3 A_1 IN_{11} \rightarrow P_4 A_3 IN_{21}$

Because $P_3 A_1 = A_2 P_1$ and $P_4 A_3 = A_4 P_2$, the rule X_4 can also be written in this way:

$X_4 : A_2 P_1 IN_{11} \rightarrow A_4 P_2 IN_{21}$

A useful image of the Agent[3] and its semantic unity can be obtained by the special *linguistic* form created by G.J. Marshal, A.

Behrooz, [76]. The semantic unity of the third order adaptive agent in the Table 2.12 and 2.15 linguistically becomes:

Activity : name of the activity for the third order agent

Environment 1 = Context 1

Environment 2 = Context 2

Environment 3 = Context 3

Environment 4 = Context 4

Complex Environment at the second order 1 = (Context 1, Context 2)

Complex Environment at the second order 2 = (Context 3, Context 4)

Objects in the context 1: variables IN_{11} and IN_{21} and methods X_1

Objects in the context 2: variables OUT_{11} and OUT_{21} and methods X_2

Objects in the context 3: variables IN_{21} and IN_{22} and methods X_3

Objects in the context 4: variables OUT_{21} and OUT_{22} and methods X_4

Aim of the adaptation channel: with P_1 and P_2 we build a channel 1 (that is the means by which the second order transformation is possible) to transform the objects 1 into the object 2, with P_3 and P_4 we build a channel 2 to transform the objects 3 into the object 4, with A_1, A_2, A_3, A_4, we build a channel 12 to transform the channel 1 into the channel 2.

Example:

When we consider the specification as the context where we operate and design the rule that we build we have:

Activity: integration redesign

Environment 1 = Environment 2 = Environment 3 = Environment 4
= Context

Object 1 : Existing object designed to a given specification

Object 2 : required object with a different specification

Object 3 : required object with a different specification

Object 4 : required object with a different specification

Aims of the adaptation channel:

(a) Given the specification and design of Object 1 and the specifi-
     cation of the object 2, we adapt the design of object 1 to give
     the design of object 2.
(b) Given the specification and design of Object 1 and the specifi-
     cation of the object 3, we adapt the design of object 1 to give
     the design of object 3.
(c) Given the specification and design of Object 3 and the specifi-
     cation and design of object 2, we integrate or adapt at the third
     order the design of object 3 and the design of object 2 to give
     the design of object 4.

In the graph in Figure 2.33 we give an illustration of the integration
redesign activity.

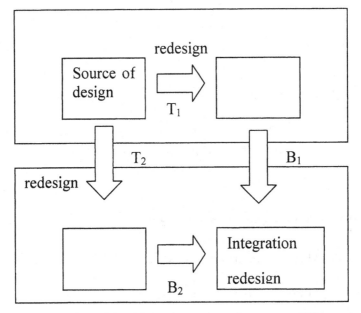

Figure 2.33. Action of the third order agent to integrate a different design and build a design for an object in the context at the convergence of B_1 and B_2.

2.5.1 Types, Transformations or Actions between the Rules for the Third Order Adaptive Agent

The Third order adaptive agent includes three adaptive agents of the second order with the same initial source N_{11}. For the three agents[2] we have three equations (2.23) which we rewrite in a simpler form:

$$P_2 X_1 = X_2 P_1, \ A_3 X_1 = X_3 A_1, \ A_2 P_1 = P_3 A_1, \tag{2.25}$$

For every equation we have the three types of transformation shown in chapter 2.4. For the *first equation* we have:

$P_1 = P_2$ and $X_1 P_1 = P_1 X_1$. In this $X_1 = X_2$

$P_1 = P_2$ and $X_1 P_1 \neq P_1 X_1$. Here we have $X_1 \approx X_2$ and X_1 is similar to X_2

$P_1 \neq P_2$ and X_1 is not similar to X_2.

When X_1 is not similar to X_2 we can use the knowledge in 2.4.3 to build the adaptive terms G_2 and H_2 for which we have:

$$P_1 = G_2 P_2 \text{ or } P_2 = P_1 H_2 \tag{2.26}$$

For (2.26) we have the compensator terms:

$X_1 \rightarrow H_1 X_1$ where $H_1 X_1$ is similar to X_2

$X_2 \rightarrow X_2 G_1$ where $X_2 G_1$ is similar to X_1

When we use the first transformation $X_1 \rightarrow H_2 X_1$ the first equation becomes:

$P_1 [H X_1] = X_2 P_1$ so P_2 may be substituted by P_1.

When we use the second transformation $X_2 \rightarrow X_2 G$ the first equation becomes:

$P_1 X_1 = [X_2 G] P_1$ so P_1 may be substituted by P_2.

For the *second equation* we have:

$A_1 = A_3$, and $X_1 A_1 = A_1 X_1$. In this case we have $X_1 = X_3$

$A_1 = A_3$, and $X_1 A_1 \neq A_1 X_1$. In this case we have $X_1 \approx X_3$ and X_1 is similar to X_3.

$A_1 \neq A_3$, X_1 is not similar to X_3.

When X_1 is not similar to X_3 we can use the knowledge in 2.4.3 to build the adaptive rules G_2 and H_2 for which we have:

$$A_1 = G_2 A_3 \text{ or } A_3 = A_1 H_2 \tag{2.27}$$

For the (2.27) we have the compensation terms:

$X_1 \to H_2 X_1$ for which $H_2 X_1$ is similar to X_2 by A_1.

$X_3 \to X_3 G_3$ for which $X_3 G_3$ is similar to X_1 by A_3.

When we use the first transformation $X_1 \to H_2 X_1$ the second equation becomes:

$A_1 [H_2 X_1] = X_3 A_1$ so A_3 may be substituted by A_1.

When we use the second transformation $X_3 \to X_3 G_3$ the second equation becomes:

$A_3 X_1 = [X_3 G_2] A_3$ so A_1 may be substituted by A_3.

Remark:

When we collect the transformations for the two cases we have the possible equations:

$$P_1 [H X_1] = X_2 P_1, A_3 X_1 = [X_3 G_2] A_3 \tag{2.28}$$

Where we change X_1 and X_3. For the third equation $A_2 P_1 = P_3 A_1$ and the (2.28) A_1 may be substituted by A_3 so the third equation becomes:

$$A_2 P_1 = P_3 A_3 \tag{2.29}$$

Because we want to substitute A_2 with A_3, we choose to change P_1 into $H_3 P_1$, so we have:

$A_3 [H_3 P_1] = P_3 A_3$ (2.30)

Where $A_3 H_3 = A_2$. A_2 is substituted by A_3. For the relations:

$P_4 A_3 X_1 = A_4 P_2 X_1 = X_4 A_2 P_1$

From the previous change that becomes:

$P_4 A_3 H_1 X_1 = A_4 P_1 H_1 X_1 = X_4 A_3 P_1$

We then have:

$P_4 A_3 = A_4 P_1$ (2.31)

Because we want to substitute A_4 with A_3, we substitute P_1 with the rule $H_4 P_1$ and then have:

$P_4 A_3 = A_3 H_4 P_1$

Where $A_3 H_4 = A_4$.

Remark:

In this way the rules A_1, A_2, A_3, A_4 are all transformed into A_3.

We show in Figure 2.34 the action of the third order agent after compensation.

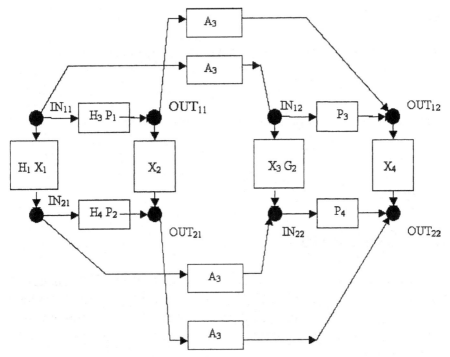

Figure 2.34. The action of the third order agent after the compensation process.

Since the first equation in (2.25) is now,

$$X_2 [H_3 P_1] = [H_4 P_1] [H_1 X_1] \qquad (2.32)$$

without the presence of the rules A_1, A_2, A_3, A_4, we can put $[H_3 P_1] = H_5 [H_4 P_1]$ so the (2.32) becomes:

$$[X_2 G_5] [H_4 P_1] = [H_4 P_1] [H_1 X_1]$$

And $X_2 \rightarrow X_2 G_5$ using the same rule becomes $[H_4 P_1]$.

Because the image of the action of adaptive agent in Table 2.13 is the action of adaptive agent in Table 2.14 the equation of which is:

$$X_4 P_3 = P_4 X_3 \tag{2.33}$$

For the previous substitution we have that (2.33) becomes:

$$X_4 P_3 = P_4 X_3 G_2 \tag{2.34}$$

When we change X_4 into $X_4 G_4$ then (2.34) becomes:

$$X_4 G_4 P_4 = P_4 X_3 G_2 \tag{2.35}$$

Where $G_4 P_4 = P_3$.

2.6 Conclusion

In conclusion the agent of the second type in Table 2.13 changes one rule into a similar rule, as shown in the image 2.14. The rules A_1, A_2, A_3, A_4 can be reduced to one only rule A_3. This transforms the action of one second order agent into another similar second order agent.

Figure 2.35 shows the final result for the action of the Third order agent.

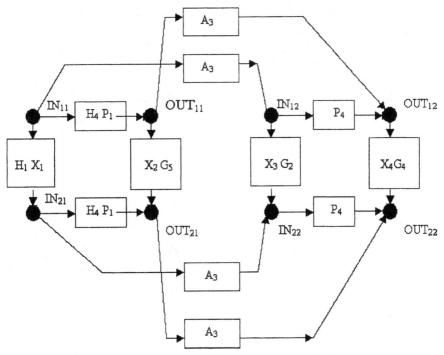

Figure 2.35. Action of the third order agent after all the compensation process for which each part is similar.

The final result can also be represented by Table 2.16 of statements.

Table 2.16. Action of the Agent[3] or semantic unity of the third order agent.

Statements	Resource	Property	Value
S_1	IN_{11}	$H_1 X_1$	IN_{21}
S_2	IN_{11}	$H_4 P_1$	OUT_{11}
S_3	IN_{11}	A_3	IN_{12}
S_4	IN_{21}	$H_4 P_1$	OUT_{21}
S_5	IN_{21}	A_3	IN_{22}
S_6	OUT_{11}	$X_2 G_5$	OUT_{21}
S_7	OUT_{11}	A_3	OUT_{12}
S_8	IN_{12}	$X_3 G_2$	IN_{22}
S_9	IN_{12}	P_4	OUT_{12}
S_{10}	IN_{22}	P_4	OUT_{22}
S_{11}	OUT_{21}	A_3	OUT_{22}
S_{12}	OUT_{12}	$X_4 G_4$	OUT_{22}

Physical Example

In the example of the secondary source in the chapter (2.4.1) in the Table 2.3b we have one transformation of the field H and the transformation of the field into the sources that generate this field. The action of the second order agent is to change the transformation of the field H into a change of the sources. We can extend this image by taking the action of one agent[3] of the third order which changes the action of the second order. This changes $s_1(x,t)$ H into $s_2(x,t)$ H, where H is the original sound field. The Agent[3] changes X_2 into X_2^* where X_2 adapts the sources to the transformation of H or $s_1(x,t)$ H and X_2^* is the result of the transformation of Agent[3] which adapts the sources to the field $s_2(x,t)$ H.

Table 2.17. Action of the Agent[3] or semantic unity of the third order agent.

Statements	Resource	Property	Value
ST_1	H	$s_1(x,t)$	s_1H
ST_2	H	P_1	P_1H
ST_3	H	$c^{-2} \partial_t$ div grad $\rho_0 \partial_t$	S_1
ST_4	s_1H	P_2	$s_2 H$
ST_5	IN_{21}	$c^{-2} \partial_t$ div grad $\rho_0 \partial_t$	S_1
ST_6	$P_1 H$	$s_2(x,t)$	$s_2 H$
ST_7	$P_1 H$	$c^{-2} \partial_t$ div grad $\rho_0 \partial_t$	S_2
ST_8	S_1	X_3	S_2
ST_9	S_1	P_3	S_3
ST_{10}	S_2	P_4	S_4
ST_{11}	$s_2 H$	$c^{-2} \partial_t$ div grad $\rho_0 \partial_t$	S_4
ST_{12}	S_3	P_4	S_4

Remark:

Actions of orders four, five and more are possible. Every action of the order n changes the action of the order (n-1) into an action of the order (n-1). We will explain the abstract structure of the nested order of actions for an adaptive agent in chapter 3. This chapter will deal with the abstract approach to equations of different orders. In Figure 2.36 we show the structure of the action of the order four.

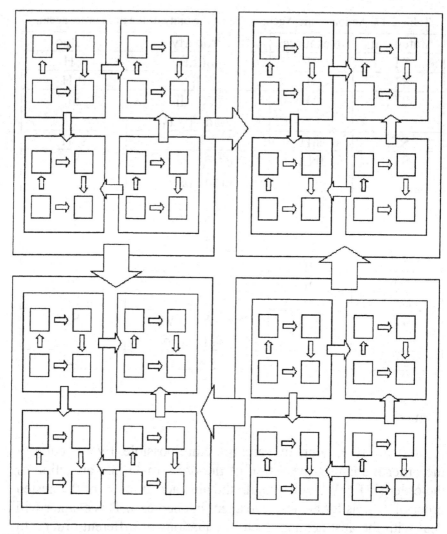

Figure 2.36. Shows the action of the fourth order agent in a nested structure of boxes and transformations or actions. In the inner box we have the simple rules. These rules transform relations at the second order.

Other rules change the box that includes boxes of the third order action The actions that change the big boxes are the fourth order actions

Bibliography

1. Penna M.P., Pessa E. and Resconi G., General System Logical Theory and its role in cognitive psychology, Third European Congress on Systems Science, Rome, 1-4 October 1996.

2. Tzvetkova G.V. and Resconi G., Network recursive structure of robot dynamics based on GSLT, European Congress on Systems Science, Rome, 1-4 October 1996.

3. Resconi G. and Tzvetkova G.V., Simulation of Dynamic Behaviour of robot manipulators by General System Logical Theory, 14-th International Symposium "Manufacturing and Robots", Lugano Switzeland, 25-27 June 1991, pp.103-106.

4. Resconi G. and Hill G., The Language of General Systems Logical Theory: a Categorical View, European Congress on Systems Science, Rome, 1-4 October 1996.

5. Rattray C., Resconi G. and Hill G., GSLT and software Development Process, Eleventh International Conference on Mathematical and Computer Modelling and Scientific Computing, Georgetown University, Washington D.C, March 31-April 3,1997.

6. Mignani R., Pessa E. and Resconi G., Commutative diagrams and tensor calculus in Riemann spaces, Il Nuovo Cimento 108B(12) December 1993.

7. Petrov A.A., Resconi G., Faglia R. and Magnani P.L., General System Logical Theory and its applications to task description for intelligent robot. In Proceeding of the sixth International Conference on Artificial Intelligence and Information Control System of Robots, Smolenize Castle, Slovakia, September 1994.

8. Minati G. and Resconi G., Detecting Meaning, European Congress on Systems Science, Rome, 1-4 October 1996.

9. Kazakov G.A. and Resconi G., Influenced Markovian Checking Processes By General System Logical Theory, International Journal General System, vol.22, pp.277-296, 1994.

10. Saunders Mac Lane, *Categories for Working Mathematician.* Springer-New York, Heidelberg Berlino, 1971.

11. Wymore A.W., *Model-Based Systems Engineering.* CRC Press, 1993.

12. Mesarovich M.D. and Takahara Y., *Foundation of General System Theory.* Academic Press, 1975.

13. Mesarovich M.D. and Takahara Y., Abstract System Theory, *In Lecture Notes in Control and Information Science 116.* Springer Verlag 1989.

14. Resconi G. and Jessel M., A General System Logical Theory, International Journal of General Systems, vol.12, pp. 159-182, 1986.

15. Resconi G. and Wymore A.W., Tricotyledon Theory of System amd General System Logical Theory, Eurocast'97, 1997.

16. Fatmi H.A., Marcer P.J., Jessel M. and Resconi G., "Theory of Cybernetics and Intelligent Machine based on Lie, Commutators", 16, 2, pp.123-164, 1990.

17. Kalman R.E., Falb P.L. and Arbib M.A., *Topics in Mathematical System Theory*, McGraw--Hill Publ, 1969.

18. Mesarovic M.D. and Takahara Y., Abstract Systems Theory, Lecture Notes in Control and Information Systems, Springer—Langer, 1989.

19. Padulo L. and Arbib M.A., *System Theory*, WB Saunders, 1974.

20. Resconi G., Rattray C. and Hill G., "The Language of General Systems Logical Theory (GSLT)", International Journal of general Systems, vol. 28, (4-5), pp.383-416, 1999.

21. Rattray C., "Identification and Recognition through Shape in Complex Systems", Computer Aided Systems Theory -- EUROCAST'95 (eds. F Pichler), 1996: R Moreno Diaz, R Albrecht), LNCS 1030, Springer—Verlag, 1996.

22. Marshall G.J. and Behrooz A., Adaptation Channels, Cybernetics and Systems 26 number 3, pp. 349-365, 1995.

23. Santilli R.M., Foundation of Theoretical Mechanics, vol I and II, Birkhoffian Generalization of Hamiltonian Mechanics, Springer Verlag, Heidelberg/NewYork, 1982.

24. James A. Crowell, Martin S. Banks, Krishna V. Shenoy and Richard A. Andersen, Visual self-motion perception during head turns, nature neuroscience, vol. 1, no. 8, pp. 732-737, 1998.

25. Array F., Norman R.Z. and Cartwright B., *Introduction à la thèorie des graphes orientés*, Dunod Paris, 1968.

26. Klir G., *Architecture of System Problem Solving*, Prenum Press, New York, London, 1985.

27. Lin Y., Development of New Theory with Generality to unify diverse disciplines of knowledge and capability of applications, Int. J. General System, vol.23, pp. 221-239, 1995.

28. Gurevich Y., Sequential Abstract State Machines Capture Sequential Algorithms, ACM Transactions on Computational Logic, vol.1 no.1, pp. 77-111, July 2000.

29. Gio Wiederhold and jan Jannink, Composing Diverse Ontologies, IFIP working on database 8th Working Conference on data base Semantic (DS-8) in Rotorna, New Zealand, 1998.

30. Chellas F., *Modal Logic: An Introduction.* Cambridge University Press, 1980.

31. Dubois D. and Parade H., *Possibility Theory.* Plemun Press, 1988.

32. Harmanec D., Klir G.J. and Resconi G., On Modal Logic Interpretation of Dempster-Shefer Theory of Evidence. Int. J. Intelligent Systems, vol. 9, pp. 941-951, 1994.

33. Hughes E. and Cresswell M.J., *An Introduction to Modal Logic.* Methuen, 1968.

34. Klir G.J. and Harmanec D., On Modal Logic Interpretation of Possibility Theory Int. J. of Uncertainty, Fuzziness and Knowledge-Based Systems, vol. 2, pp. 237-245, 1994.

35. Klir G.J. and Yuan B., *Fuzzy Sets and Fuzzy Logic.* Prentice Hall, 1995.

36. Murai T., Nakata M. and Shimbo M., Ambiguity, Inconsistency, and Possible-Worlds: A New Logical Approach. Proceedings of the Ninth Conference on Intelligence Technologies in Human-Related Sciences, Leòn, Spain, pp. 177-184, 1996.

37. Murai T., Kanemitsu H. and Shimbo M., Fuzzy Sets and Binary-Proximity-based Rough Sets, Information Science, vol. 104, pp. 49-80, 1998.

38. Resconi G., Klir G.J. and St. Clair U., Hierarchical Uncertainty Metatheory Based Upon Modal Logic, Int. J. of General Systems, vol. 21, pp. 23-50, 1992.

39. Resconi G., Klir G.J., St. Clair U. and Harmanec D., On the Integration of Uncertainty Theories. Int. J. of Uncertainty, Fuzziness, and Knowledge-Based Systems, vol. 1, pp. 1-18, 1993.

40. Resconi G. and Rovetta R., Fuzzy Sets and Evidence Theory in a Metatheory Based Upon Modal Logic, Quaderni del Seminario Matematico di Brescia, n.5, 1993.

41. Resconi G., Klir G.J., Harmanec D. and St. Clair U., Interpretations of Various Uncertainty Theories Using Models of Modal Logics: A Summary, Fuzzy Sets and Systems, vol. 80, pp. 7-14, 1996.

42. Resconi G. and Murai T., Field Theory and Modal Logic by Semantic Field to Make Uncertainty Emerge from Information, Int.J.General System, 2000.

43. Shafer G., *A Mathematical Theory of Evidence*, Princeton University Press, 1976.

44. Sowa J.F., *Knowledge Representation: Logic, Philosophical, and Computational Foundation.* PWS Publishing Co., Pacific Grove, CA., 1999.

45. Türkşen I.B., Theories of Set and Logic with Crisp and Fuzzy Information Granules, J. of Advanced Computational Intelligence (to appear in the Special Issue), 2000.

46. Türkşen I.B., Computing with Descriptive and Veristic Words, Proceedings of NAFIPS'99 (Special Invited Presentation), New York, pp. 13-17, June 10-12, 1999.

47. Türkşen I.B., Fuzzy Normal Forms, FSS, pp. 253-266, 1994.

48. Türkşen I.B., Interval-Valued Fuzzy Sets and 'Compensatory AND', FSS, pp. 295-307, 1994.

49. Türkşen I.B., Non-Specificity and Interval-Valued Fuzzy Sets, FSS, pp. 87-100, 1995.

50. Türkşen I.B., Interval-Valued Fuzzy Sets Based of Normal Forms, FSS, pp. 191-210, 1986.

51. Zadeh L.A., 'Fuzzy Sets, Information and Control, vol. 8, pp. 338-353. 1965.

52. Zadeh L.A., A Simple View on the Dempster-Shafer Theory of Evidence and Its Implication for the Rule of Combination, AI Magazine, vol. 7, pp.85-90, 1986.

53. Zimmerman H.J. and Zysno P., Latent Connectives in Human Decision Making, FSS, pp. 37-51, 1980.

54. McCullough W. and Pitts W.H., "A logical calculus of the ideas immanent in nervous activity", Bull. Math. Biophys. 5, pp. 115-133, 1943.

55. Murre J.M.J., *Learning and categorization in modular neural networks*, Erlbaum, Hillsdale, NJ, 1992.

56. Koch C., Computation and the single neuron, Nature vol.385, no. 16, pp. 207-210, January 1997.

57. Benjafield J.G., *Cognition*. Prentice-Hall, Englewood Cliffs, NJ, 1992.

58. Resconi G., "The morphogenetic Neuron" in Computational Intelligence : Soft Computing and Fuzzy.- Neuro Integration

with Application, Springer NATO ASI Series F Computer and System Science vol 162, pp. 304-331, 1998. editors Okyay Kaynak, Lotfi Zadeh, Burhan Turksen, Imre J.Rudas.

59. Resconi G., Pessa E. and Poluzzi R., "The Morphogenetic Neuron", Proceedings fourteenth European meeting on cybernetics and systems research, pp.628-633, April 14-17, 1998.

60. Resconi G. and van der Wal A.J., A data model for the Morphogenetic Neuron,Int. J. General System, vol.29, no.1, pp.141-174, 2000.

61. Resconi G. and van der Wal A.J., Single neuron as a quantum computer-morphogenetic neuron, Intelligent techniques and Soft Computing in Nuclear Science and Engineering, World Scientific, Proceeding of the 4th International FLINS Conference Bruges, Belgium, August 28-30, 2000.

62. Barenco A., Quantum Computation, Thesis Lincoln College University of Oxford Trinity Term, 1996.

63. Shor P.W., Scheme for reducing decoherence in quantum computer memory, Phys.Rev. A52, pp. 2493-2496, 1995.

64. Nobuo S., "The extent to which Biosonar information is represented in the Bat Auditory Cortex", in: "Dynamic Aspects of Neocortical Function", ed. G.M. Edelman, John Wiley, New York, 1984.

65. Salinas E. and Abbott L.F., "A model of multiplicative neural responses in parietal cortex", Proc. Natl. Acad. Sci. USA, vol.93, pp.11956-11961, 1996.

66. Salinas E. and Abbott L.F., "Invariant Visual responses From Attentional gain Fields", The American Physiological Society, pp.3267-3272, 1997.

67. Borisenko A.I. and Tarapov I.E., "Vector and Tensor analysis with applications", Prentice-Hall, 1968.

68. van der Wal A.J., The role of fuzzy set theory in the conceptual foundations of quantum mechanics: an early application of fuzzy measures in "Foundations and applications of possibility theory, vol 8 of "Advances in Fuzzy Systems", World Scientific, Singapore, FAPT 95, pp.234-245, 1995.

69. Resconi G., George G, Klir and Pessa E., Conceptual foundation of quantum mechanics: the role of evidence theory, quantum sets, and modal logic, International Journal of modern Physics C, vol.10, no.1, pp.29-62, 1999.

70. Shafer G., *A mathematical theory of evidence*, Princeton University Press, 1976.

71. van der Wal A.J., Sensor synergetics: the design philosophy of the sensor fusion demonstrator and testbed at TNO-FEL,31,pp 1-26 in "Multi-sensor systems and data fusion for telecommunications, remote sensing and radar", AGARD-CP-595, Lisbon, ISBN 92-836-0051-17, 1998.

72. Jonathan D. Walls, Kathaleen C. Anderson and Earl K. Miller, Single neurons in prefrontal cortex encode abstract rules, Nature vol.41/21, June 2001.

73. Alexandre Pouget and Lawrence H. Snyder, Computational approaches to sensorimotor transformation supplement to nature neuroscience, vol. 3, pp.1192-1198, supplement November 2000.

74. Maximilian Riesenhuber and Tomaso Poggio, Models of object recognition, supplement to nature neuroscience vol.3, pp.1199-1204, supplement Novembre 2000.

75. Mariusz Nowostawki, Martin Puvis and Stephen Cranefield, A layered Approach for Modelling Agent Conversations, The information Science Discussion paper series, Number 2001/05, March 2001, ISSN 1172-6024.

76. Marshal G.J. and Behrooz A., Adaptive Channels, Cybernetic and Systems, vol.26, no.3, pp349, 1995.

77. Jessel M., Acoustique Théorique, Masson et Cie Editeurs, 1973.

78. Wolfgang Pauli, Relativitatstheorie Encyklopadie der mathematishen Wissenschaften, vol.5 article 19 Teubner Lipsia 1921.

Chapter Three

Abstract Theory of Adaptive Agents

3.1 Introduction

The syntactic and semantic aspects of the formal language of General System Logical Theory (GSLT), also known as Abstract Theory of Adaptive Agents is presented in order to improve the formal description of the action of the adaptive agents. Following an introduction to the language of GSLT, we show the possibility of the action of the Adaptive Agents in a number of areas. The category theory will be the abstract structure to model the adaptive agent at different orders. Every semantic unity which describes an action of the adaptive agent is expressed by an elementary unity denoted Elementary Logical Systems (ELS) which are the basic components of GSLT. GSLT uses the input-output paradigm to represent any action. The static classical form of categorical structures is transformed as a dynamic transformation of objects. The input-output paradigm of system theory is well known and is extensively developed in the literature.

Our work is inspired by that of Mesarovic and Takahara on the mathematical foundation of system theory (Mesarovich and Takahara, 1975 [11]) and the subsequent presentation of abstract system theory. General System Logical Theory is based on the input-output paradigm and was first proposed by Resconi and Jessel (Resconi and Jessel, 1986) [19]. It has since been applied to deriving the basic rules of tensor analysis in non- Euclidean space (Mignani et al., 1993) [12]. The study of robots, Penna et al., 1996 [14], Resconi and Tzvetkova, 1991[20]: Petrov et al., 1994) [15].

Extensions of Markov Chains (Kazakov and Resconi, 1994) [6], General Relativity (Mignani et al., 1993) [12], and the dynamic control of sound waves. Other areas considered included the work of Beth (1993) [1] on Cryptology, Hill (1994) [4] on Configuration of Complex Systems and of Fatmi et al. (1990) [2] on Intelligent Machines.

In this chapter we introduce the language of GSLT as the formal description of adaptive agents at different orders. We show the power of the GSLT language and the language of the Adaptive Agent by using it to express some well-known Category Theory Concepts (MacLane, 1971) [8]. Using the language of GSLT, and guided by the graphic representation of the action, we are led to a particular n-dimensional diagram generated by a simple recursive rule. In this way we simplify the categorical representation and are able to overcome certain difficulties that appear when attempting to introduce a hierarchical structure within a categorical framework. We note certain limitation in going from morphism (where morphism is an arrow between objects in one category) to functor (where functor is a transformation of one category into another) to natural transformation (the natural transformation changes one functor into another). Consequently we recognise the inherent difficulties in going beyond these three orders of hierarchical structure (Resconi and Hill, 1996) [21].

Category theory has provided us with a number of unifying concepts of significance but it is poorly suited for practical applications in which the input output paradigm is standard. The general properties and constructions provided by category theory clearly impinge on the development of complex real systems. It seems important, then, to provide some means of interfacing the power of category for use in practical system development.

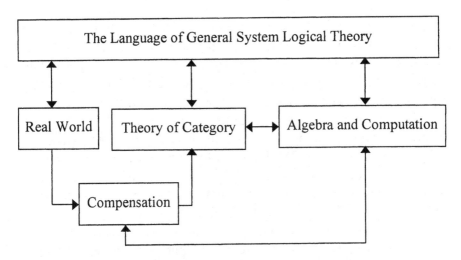

Figure 3.1. GSLT Language Bridge which Spans the Real World, Category Theory and Algebra.

The role of GSLT in problem development and solution hinges on two notions. Firstly, "compensation" provides an intuitive basis for the constructive nature of categorical concepts. It also provides an important connection to the notion of compensation prevalent in physical systems. Secondly, the link with reality for GSLT derives by identifying the "intrinsic equation" generated from the internal structure of Elementary Logical Systems (ELS) or action defined in the chapter 2. As Figure 3.1 illustrates, the language of GSLT is interpreted in the real world, in the categorical world, and in the algebraic and computational world. The real world is dynamic and time-related; the categorical world is static. The GSLT language, with its input-output paradigm, enables the classical static form of Category Theory to be transformed into a dynamic form consisting of objects, morphisms, functors, natural transformations and other higher order entities. We remember that every mathematical concept can be represented by the conceptual theory of the Category. So we compare actions at a different order as expressed by ELS in the GSLT with static mathematical structures as algebra, geometry, analysis and so on. Using the GSLT language representation, the categorical world better matches the real world. In Fig.3.1, we show a connection between "compensation" and algebra. In fact, compensation will be represented by compositions in a particular algebra. Category theory is expressed in category algebra. Algebra is part of the language of the GSLT and its semantic elements. In Chapter Two we have shown that every action at order two, three or more is controlled by a set of equations among the operators within an algebra. The GSLT shows for the first time the connection between the action and the algebra that controls the action. Compensation represents the information, the knowledge and the activity required to neutralise the difference between the true solution in an all-knowing real world and the actual solution in a real world subject to partial information and partial knowledge. Compensation integrates the partial knowledge on the rules in different contexts in such a way as to establish a coherent connection between rules in different contexts at different orders. Compensation can integrate the knowledge so as to have a relation of similarity among different rules in different contexts or among actions with a high order as discussed in Chapter 2.

The aim of this chapter is to present the language of GSLT that is the abstract language of the action of adaptive agents of a different order. Later in the chapter, the power of the language of the GSLT or its adaptation in different areas is highlighted. There is an interesting connection between the General System Problem Solver (GSPS) and GSLT. In fact, Klir (1985) [7] describes GSPS by source systems. That is to say a source in one respect exemplified by scientific investigations is a source of empirical data. That is a source of abstract images expressed in the GSPS language of some real-world phenomena. Data systems are given by a connecting pair of a source system and specific variables in the data systems. We may study instances of measures for given support systems such as time and space. Generative systems and Meta -Systems open up the possibility of different orders of systems of high logical and categorical complexity. All knowledge within GSPS can be used within the GSLT to build different models in the formal GSLT language. The two languages are complementary but GSLT is more formal and general while GSPS is more oriented to problem solving.

3.2 The Syntactic Structure K of the Language of GSLT

To define a grammar we need:

> a set of terminal symbols   : Σ
> a set of non terminal symbols : V
> a start symbol   : T
> a set of production rules   : P

Where Σ generates the set of words that appear as words of the language L generated by grammar. The symbols in Σ must contain all the symbols that appear in the words of the language:

$$\Sigma = \{\, ob_1, ob_2, ..., \; op_1, op_2, ... \rightarrow, + \,)$$

V is a set of symbols, $V \cap \Sigma = \varnothing$. These symbols are needed to write the grammar rules. Therefore, no symbol in V may appear in the language L

$$V = \{\, A^0, A^1, ..., A^{p-1}, A^p, F^1; F^2, ..., F^p, T \,\}$$

$T \in V$ is the start symbol and all words of the language L are generated from T by the grammar.

A production rule has the form $\alpha \Rightarrow \beta$, where $\alpha, \beta \in (V \cup \Sigma)$ *, and ( )* is the set of all strings or words over the symbols $V \cup \Sigma$ using the operation of concatenations (Manna, 1974)[9].

3.2.1 Syntactic Structure of GSLT of the Order p

The alphabet Σ of the GSLT language known as K_n, is:

$$\Sigma = \{ ob_1, ob_2, ..., op_1, op_2, ... \to, + \ ).$$

Where ob_1, ob_2, ... form a finite or denumerable set of symbols and the op_1, op_2,.... form another finite or denumerable set of symbols. Symbol "$\to$ "denotes a "connector" and A^k an Elementary Logical System of the order k or ELS(k) or Agent Action at the Order k. The set of variables is:

$$V(p) = \{ A^0, A^1, ..., A^{p-1}, A^p, F^1; F^2, ..., F^p, T \} \qquad (3.1)$$

The set of variables depends on $p \in N$. That is different sets of variables exist for different values of p. Symbol T is the start symbol. This is a non-terminal symbol.

Symbol "$\to$ " is a "connector", A^k is Elementary Logical System of the order k or ELS(k), F^k are hyper-functions of order k where k=1, 2, ..., p.

Examples

$$V(1) = \{A^0, A^1, F^1, T \}$$

$$V(2) = \{A^0, A^1, A^2, F^1, F^2, T \}$$

The production rules used for the formal GSLT language of the order p are:

R1: Specification rule

$$T \Rightarrow A^p$$
$$A^p \Rightarrow A^{p-1} \to F^p \to A^{p-1}, \quad p \geq 1 \qquad (3.2)$$

We shall refer to $A^{p-1} \to F^p \to A^{p-1}$ as a one- dimensional word. From one-dimensional words we proceed to two dimensional words:

$$A^{p-1} \to F^p \to A^{p-1} \quad \Rightarrow \quad
\begin{array}{ccc}
A^{p-2} \to F^{p-1} \to A^{p-2} \\
\downarrow \qquad\qquad \downarrow \\
F^{p-1} \qquad\quad F^{p-1} \\
\downarrow \qquad\qquad \downarrow \\
A^{p-2} \to F^{p-1} \to A^{p-2}
\end{array} \qquad (3.3)$$

Three dimensional words are obtained from two dimensional words as follows:

$$(3.4)$$

From the three dimensional words we can obtain four dimensional words and words of higher order until we obtain p-dimensional words. Here the only symbols present are $\{\to, A^0, F^1\}$. The specification rule increases the complexity of the structure that connects different types of symbols, and at the same time reduces the set of symbols used. At the end of the process only the symbols $\{\to, A^0, F^1\}$ are used.

Notation definitions

In the diagrams associated with an Elementary Logical System there exist only one source or focal point and one sink point.

Example:

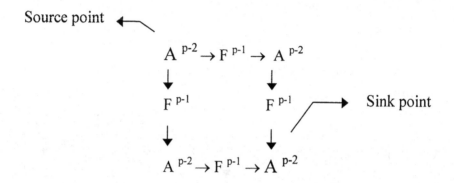

Source point

$$A^{p-2} \to F^{p-1} \to A^{p-2}$$
$$\downarrow \qquad\qquad \downarrow$$
$$F^{p-1} \qquad\qquad F^{p-1} \qquad \longrightarrow \quad \text{Sink point}$$
$$\downarrow \qquad\qquad \downarrow$$
$$A^{p-2} \to F^{p-1} \to A^{p-2}$$

R2: Sequential composition

$$A^{p-1} \to F^{p} \to A^{p-1} + A^{p-1} \to F^{p} \to A^{p-1} \;\Rightarrow\; A^{p-1} \to F^{p} \to A^{p-1}$$
$$\to F^{p} \to A^{q-1} \tag{3.5}$$

$$A^{p-1} \to F^{p} \to A^{p-1} + \quad A^{p-1} \to F^{p} \to \quad A^{p-1} \quad \Rightarrow$$
$$\downarrow \qquad\qquad \downarrow$$
$$F^{p} \qquad\qquad F^{p}$$
$$\downarrow \qquad\qquad \downarrow$$
$$A^{p-1} \to F^{p} \to A^{p-1}$$

$$A^{p-1} \to F^p \to A^{p-1} \to F^p \to A^{p-1}$$
$$\downarrow \qquad\qquad\qquad \downarrow$$
$$F^p \qquad\qquad\qquad F^p$$
$$\downarrow \qquad\qquad\qquad \downarrow \qquad\qquad (3.6)$$
$$A^{p-1} \to F^p \to A^{p-1}$$

We show only a few of the possible sequential composition rules, but we can extend this rule to any specification of an elementary logical system.

R3: Generation rule

Using the generator ELS, $A^p = A^{p-1} \to F^p \to A^{p-1,}$ and the generator $A^p = A^{p-1} \to F^p \to A^{p-1}$, we can generate the following relationships

$$A^{p-1} \to F^p \to A^{p-1}$$
$$\downarrow \qquad\qquad \downarrow$$
$$F^p \qquad\qquad F^p$$
$$\downarrow \qquad\qquad \downarrow \qquad\qquad (3.7)$$
$$A^{p-1} \to F^p \to A^{p-1}$$

Other examples of the generation rule are shown in Equations (3.38) and (3.39).

R4: Up rule or extension

$$A^{p-1} \to F^p \to A^{p-1}$$
$$\downarrow \qquad\qquad \downarrow$$
$$A^{p-1} \to F^p \to A^{p-1} \Rightarrow \quad F^p \qquad\qquad F^p \qquad\qquad (3.8)$$
$$\downarrow \qquad\qquad \downarrow$$
$$A^{p-1} \to F^p \to A^{p-1}$$

From the ELS $A^P = A^{P-1} \rightarrow F^P \rightarrow A^{P-1}$ we can obtain:

$$A^{P+1} = A^P \rightarrow F^{P+1} \rightarrow A^P \tag{3.9}$$

and by using the specification rule, the two-dimensional word for A^{P+1} can be generated as:

$$A^P \rightarrow F^{P+1} \rightarrow A^P \quad \Rightarrow \quad \begin{array}{ccc} A^{P-1} \rightarrow F^P \rightarrow A^{P-1} \\ \downarrow \qquad\qquad \downarrow \\ A^P \qquad\qquad A^P \\ \downarrow \qquad\qquad \downarrow \\ A^{P-1} \rightarrow F^P \rightarrow A^{P-1} \end{array} \tag{3.10}$$

R5: Down rule or Projection

The Down rule is the reverse of the Up rule in R4.

$$\begin{array}{ccc} A^{P-1} \rightarrow F^P \rightarrow A^{P-1} \\ \downarrow \qquad\qquad \downarrow \\ F^P \qquad\qquad F^P \\ \downarrow \qquad\qquad \downarrow \\ A^{P-1} \rightarrow F^P \rightarrow A^{P-1} \end{array} \quad \Rightarrow A^{P-1} \rightarrow F^P \rightarrow A^{P-1} \tag{3.11}$$

One part of the word in the two-dimensional form is projected onto one word in one dimension.

R6: Terminal rules

$$A^0 \Rightarrow ob_1, ob_2, ...$$

$$F^1 \Rightarrow op_1, op_2, ...$$

Where, ob_k, op_k $k = 1,2....$ are elements of the alphabet Σ.

Example of GSLT language

Alphabet $\Sigma := \{ob_1, ob_2, ob_3, ob_4, op_1, op_2, op_3, op_4, \rightarrow, +\}$

Variables $V(2) = \{A^0, A^1, A^2, F^1, F^2, T\}$

R1: Specification

$$\mathbf{T} \Rightarrow \mathbf{A}^2$$
$$\mathbf{A}^2 \Rightarrow \mathbf{A}^1 \rightarrow \mathbf{F}^2 \rightarrow \mathbf{A}^1 \qquad (3.12)$$

Where, $A^1 \rightarrow F^2 \rightarrow A^1$ is one-dimensional word.

From the one-dimensional words a two-dimensional word is formed.

$$
A^1 \rightarrow F^2 \rightarrow A^1 \quad \Rightarrow \quad
\begin{array}{ccc}
A^0 \rightarrow F^1 \rightarrow A^0 \\
\downarrow \qquad\qquad \downarrow \\
F^1 \qquad\qquad F^1 \\
\downarrow \qquad\qquad \downarrow \\
A^0 \rightarrow F^1 \rightarrow A^0
\end{array}
\qquad (3.13)
$$

R2: Sequential composition

$$A^1 \rightarrow F^2 \rightarrow A^1 + A^1 \rightarrow F^2 \rightarrow A^1 \Rightarrow A^1 \rightarrow F^2 \rightarrow A^1 \rightarrow F^2 \rightarrow A^1 \qquad (3.14)$$

R3: Generation

From $A^0 \rightarrow F^1 \rightarrow A^0$ and $A^0 \rightarrow F^1 \rightarrow A^0$ we generate:

$$A^0 \to F^1 \to A^0$$

$$\downarrow \qquad\qquad \downarrow$$

$$F^1 \qquad\qquad F^1$$

$$\downarrow \qquad\qquad \downarrow \qquad\qquad\qquad\qquad (3.15)$$

$$A^0 \to F^1 \to A^0$$

R4: Up rule

$$A^1 = A^0 \to F^1 \to A^0 \quad\Rightarrow\quad \begin{array}{c} A^0 \to F^1 \to A^0 \\ \downarrow \qquad\qquad \downarrow \\ F^1 \qquad\qquad F^1 = A^2 \\ \downarrow \qquad\qquad \downarrow \\ A^0 \to F^1 \to A^0 \end{array} \qquad (3.16)$$

R5: Down rule or Projection

The Down rule is the reverse of the Up rule in R4.

$$A^2 = \begin{array}{c} A^0 \to F^1 \to A^0 \\ \downarrow \qquad\qquad \downarrow \\ F^1 \qquad\qquad F^1 \\ \downarrow \qquad\qquad \downarrow \\ A^0 \to F^1 \to A^0 \end{array} \Rightarrow A^0 \to F^1 \to A^0 = A^1 \qquad (3.17)$$

One part of the word in two dimensions is projected onto one word in one-dimension.

R6 : Terminal rules

$$A^0 \Rightarrow ob_1, ob_2, ob_3, ob_4$$

$$F^1 \Rightarrow op_1, op_2, .op_1, op_2 \qquad (3.18)$$

On substituting the terminal symbols in the two-dimensional word we obtain:

$$A^2 \Rightarrow
\begin{array}{ccc}
ob_1 \to op_1 \to ob_2 \\
\downarrow \qquad\qquad \downarrow \\
op_2 \qquad\qquad op_4 \\
\downarrow \qquad\qquad \downarrow \\
ob_3 \to op_3 \to ob_4
\end{array}
\qquad (3.19)$$

3.3 The Semantics of the Language of GSLT

We will now study the **semantic structure** of the GSLT.

Given a mathematical structure M of objects, operators and relations, and a function C between the objects, operators and relations in M and terminal symbols in the **syntactic structure K** of GSLT, when all the productions in K are **valid** in the structure M with respect to correspondence C we say that M is a *Model* of K with respect to C.

We note that all the productions in K are **valid** in the structure M when, for all i and j, the two consecutive operators op_k and op_h in M can be composed.

For all k, h in the chain:

$$\text{ob}_i \rightarrow \text{op}_k \rightarrow \text{ob}_{i+1} \rightarrow \text{op}_h \rightarrow \text{ob}_{i+2} \tag{3.20}$$

There exists for the model M an operator " $\bullet$ " for which:
$$\text{op}_{k,h} = \text{op}_k \bullet \text{op}_h \tag{3.21}$$

The word "composition" is represented by the formula (3.21).

For example we generally have in mathematics as:

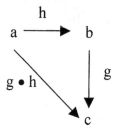

The next section discusses various aspects of a given language L_K for diagrams of particular shapes generated by the grammar K. We have a model M, and an algebraic structure having objects, operators and relations. The correspondence C maps the terminal symbols of K into corresponding objects and operators in M. That is an "ob" as a terminal symbol of K is mapped to an object of M and an "op" as a terminal symbol of K is mapped to an operator of M. This is done so that the word in L_K (a diagram of a particular shape) has the "same shape" in terms of objects and operators of M.

The question arises: how does C extend as a mapping from terminals of K to objects and operators in M to a mapping in L_K to the "same shape" diagram in M? This seems to require that L_K has an algebraic structure of the same kind as M.

Now, a production in K is valid in M with respect to C whenever chains of operations in M are comparable. This is true for " same shape" diagrams in M showing sequences of operators from M. The mathematical description that we are using here is very similar to the algebraic descriptions of language introduced by Teodor Rus (University of Iowa) and Charles Rattray, in the 1970s and 1980s. Teodor Rus (1991) [22] has extended this algebraic description of languages further, and has developed a new technology for the design and implementation of full-scale translator generators (Rus and Halverson 1994) [23]. Teodor Rus and Charles Rattray (1976)[17] based the algebraic theory of language on algebras with schemes of operators (Higgins, 1963 and 64) [3], which were mainly concerned with context- free languages used for programming languages. The introduction of relations into (partial) algebras with schemes of operators (known as (Sigma- algebras) was a simple extension (Rattray, 1975)[16].

3.3.1 Model of GSLT and Action in Adaptive Agents

In chapter 2 we discussed many different types of actions at the first , second and higher orders. These are actions of an Adaptive Agent. In the second order we have the graphic representation of the action and it is repeated here for convenience.

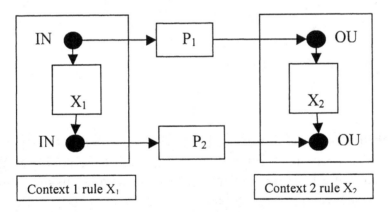

Figure 3.2. A description of the semantic unity for the second order agent that changes the rule X_1 within the context 1 into the rule X_2 within the context 2.

Using the abstract language of the GSLT the previous diagram can be represented in an abstract way by this diagram:

$$A^0 \to F^1 \to A^0$$
$$\downarrow \qquad\qquad \downarrow$$
$$A^2 \Rightarrow \qquad F^1 \qquad\qquad F^1 \qquad\qquad\qquad (3.22)$$
$$\downarrow \qquad\qquad \downarrow$$
$$A^0 \to F^1 \to A^0$$

Where A^0 are the objects (IN_1 , IN_2 , OUT_1, OUT_2) , the four functions F^1 are the rules (X_1 , X_2 , P_1, P_2). The four elements of the type:

$$A^0 \to F^1 \to A^0$$

which are the four sides, two vertical and two horizontal , of this ELS(2):

$$A^0 \to F^1 \to A^0$$
$$\downarrow \qquad\qquad \downarrow$$
$$F^1 \qquad\qquad F^1$$
$$\downarrow \qquad\qquad \downarrow$$
$$A^0 \to F^1 \to A^0$$

are all ELS of the first order or the four rules with the domain and range. The structure in Figure 3.2 is a model of the universal abstract structure in (3.22) that represents a semantic abstract unity denoted ELS at the second order.

3.3.2 Logic Predicate of Coherence and Model M of GSLT

Let us consider the two-dimensional word:

$$ob_1 \rightarrow op_1 \rightarrow ob_2$$
$$\downarrow \qquad\qquad \downarrow$$
$$op_2 \qquad\qquad op_4 \qquad\qquad\qquad (3.23)$$
$$\downarrow \qquad\qquad \downarrow$$
$$ob_3 \rightarrow op_3 \rightarrow ob_4$$

For model M:

$$C = [ob_1 \rightarrow x_1 , ob_2 \rightarrow x_2 , ob_3 \rightarrow x_3 , ob_4 \rightarrow (x_4 , x_5), op_1 \rightarrow A_1 ,$$
$$op_2 \rightarrow A_2 , op_3 \rightarrow A_3 , op_4 \rightarrow A_4]$$

where in M , x_4 is obtained from x_1 and from the operators A_2, A_3, and x_5 is obtained from x_1 and from the operators A_1 , A_4 . In (3.23), with the sink object ob_4, we associate the elements x_4 and x_5 in the model M. We introduce a logic predicative form inside the two-dimensional word (3.23):

$$A (x_4 , x_5) \in \{True , False\} ;$$

the predicate can have any form, for example:

$$A (x_4 , x_5) \equiv (x_4 = x_5) \qquad\qquad\qquad (3.24)$$

This gives us the relationship within the Elementary Logical System in (3.23).

At any ELS object, with which we can associate the set $(x_1 , x_2 ,.....,x_n)$ we can define the logic predicate:

$$A (x_1, x_2, \ldots, x_n) \in \{True, False\} \tag{3.25}$$

Part of the General System Logical Theory name came from the possibility of introducing logic predicates inside the Elementary Logical System.

3.3.3 Definition of Coherent Model

Model M is a **coherent model** when relations in M are closed predicates:

$$\forall (x_1, x_2, \ldots, x_p) \, A \, (x_1, x_2, \ldots, x_p) = True \tag{3.26}$$

The predicate A is True for all the objects of A^P in the model M. In all the other cases the model M is incoherent. In chapter 2 the coherent equations in a different order of adaptive agent actions are the logic predicates of coherence. For example in Figure 3.2 the logic predicate of the coherence is:

$$X_2 P_1 = P_2 X_1 \tag{3.27}$$

3.3.4 Model M of K and Coherent Equations

When the predicate form in (3.25) is a set of equations in the model M:

$$A (x_1, x_2, \ldots, x_p) \equiv (x_1 = x_2 = \ldots \ldots = x_p) \tag{3.28}$$

where $x_1, \ldots, x_p$ are elements in the model M. The predicate form of (3.28) contains intrinsic equations.

3.4 Compensation Functions

If the model M is not coherent, then it can be changed so that re-sulting new model M' will be coherent. The operation by which we transform a non-coherent model into a coherent model is compen-sation. Let's consider the rule R4.

For p = 2, R4, R1 and the terminal rule generate the words. If p =2 is substituted in (3.16) the resulting diagram consists of the ele-ments A^1 and F^2. The terminal rules do not apply here. We need therefore to apply R1 also to obtain (3.4).

A partial model M for (3.4) has:

C : ($ob_1 \rightarrow x_1$, $ob_2 \rightarrow x_2$, $ob_3 \rightarrow x_3$, $ob_1 \rightarrow$ (x_4, x_5) , $op_1 \rightarrow T$, $op_2 \rightarrow A$,$op_3 \rightarrow T$,$op_4 \rightarrow B$)

where x_1 , x_2 , x_3 , x_4 are n-dimensional vectors on real numbers and T , B1 , B2 are matrices that transform vectors. With model C we obtain the A^2:

$$
\begin{array}{ccc}
x_1 \rightarrow & T & \rightarrow x_2 \\
\downarrow & & \downarrow \\
B1 & & B2 \\
\downarrow & & \downarrow \\
x_3 \rightarrow & T \rightarrow & (x_4, x_5)
\end{array}
\qquad (3.29)
$$

where $x_4 = T\ B1\ x_1$, $x_5 = B2\ T\ x_1$.

When the predicate is P (x_4 , x_5) $\equiv x_4 = x_5$, then P can be false and the model M can be incoherent. However, from (3.4) we see that a two-dimensional word A^2 gives rise to a three -dimensional word

A^3. The extra dimension reveals many more degrees of freedom in the structure and by suitable choice of mapping C we can hope to generate a coherent model M'. The three- dimensional word A^3 can be viewed as a transformation of a two-dimensional word into a different two-dimensional word by means of a suitable choice of operators op_6, op_5, op_7 and op_8.

The diagrams (3.4) suggest that a 2-dimensional word can have an associated 3-dimensional word. Any model giving the semantics for the language must give meaning to both words. We may have a "partial" model M which relates only to the left-hand word. That is, a model M where some of the elements of the model mapping C are not yet fixed.

Model M' is such that the model mapping C' can give meaning to the 3-dimensional word. It should also give meaning to the 2-dimensional word as well. Then:

$$C' \supseteq C \cup (\ op_6 \rightarrow \lambda_1\ ,\ op_5 \rightarrow \lambda_2\ , op_7 \rightarrow \lambda_3\ , op_8 \rightarrow \lambda_4\ )$$

That is by means of forms (3.4), (3.29) and model M, the right hand side of (3.4) may be viewed as:

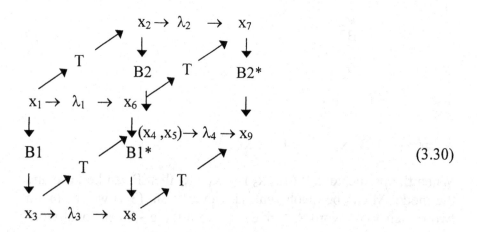

$$(3.30)$$

In the diagram (3.31) it is possible to see the transformation of diagram (3.29) into (3.31):

$$\lambda_1 x_1 \rightarrow \quad T \quad \rightarrow \quad \lambda_2 x_2$$
$$\downarrow \qquad\qquad\qquad \downarrow$$
$$B1 \qquad\qquad\quad B^*$$
$$\downarrow \qquad\qquad\qquad \downarrow$$
$$\lambda_3 x_3 \rightarrow T \rightarrow \lambda_4 (x_4, x_5) = (x_9, x_9) \qquad\qquad\qquad (3.31)$$

Thus, in the model M' (3.31), we have $\lambda_1 x_1 = x_6$, $\lambda_2 x_2 = x_7$, $\lambda_3 x_3 = x_8$,.with λ_1 , λ_2, λ_3 matrices. For the sink object ob_8 we have:

$$\lambda_4 (x_4, x_5) = (x_9, x_9)$$

and M' is a coherent model. Putting $\lambda_1 = \lambda_2 = \lambda_3 = $ Identity and $\lambda_4 = \lambda$, for the coherence of M', we must have:

$$T \, B1 \, x_1 = \lambda \, B2 \, T \, x_1 = B2^* \, T \, x_1$$

Where,

$\lambda \, B2 = B2^*$ and for coherence $\lambda \, B2 \, T = T \, B1$

and,

$$\lambda \, B2 = (T \, B1) (B2 \, T)^{-1} = T \, B1 \, T^{-1}$$

In many cases it is useful to put:

$$\lambda \, B2 \, T = B1 \, T + \eta = TB1$$

where, $\eta = TB1 - B1T$.

We then obtain,

$\lambda\ B2 = B1 + [\ TB1 - B1T\ ]\ T^{-1}$

where $[\ TB1 - B1T\ ]$ is the commutator operator between T and B1. The conclusion is that using a three-dimensional word (3.30) we can calculate the operator $\lambda\ B2$ or λ. By this means the non-coherent model M (3.29) is transformed into a coherent model M' (3.31).

A Remark concerning Algebra:

The compensation operation is always related to operations between objects in the model M'. Before using any compensation calculus, it is necessary to define the operations to be used within the Elementary Logical System. As stated in the introduction, compensation is related to the algebraic structure fundamentals for any calculus.

Given A^{p-1} with an incoherent model M, it is possible to find the element F^n by which using A^{p-1} (model M) we obtain another A^{p-1} whose model M' is coherent. Symbolically,

$A^p = A^{p-1} \rightarrow F^p \rightarrow A^{p-1}$

The A^p *transforms an incoherent ELS into a coherent one.* For the ELS at the second order (in Figure 3.2), we have a first compensation when we change the output rule X_2 into the rule $X_2\ G$ so that the (3.27) can be written as:

$X_2\ G\ P_2 = P_2\ X_1$ (3.32)

Where $G\ P_2$ is equal to P_1, in (3.32) and in Chapter 2 X_2 is not similar to X_1. However $X_2\ G$ is similar to X_1 by P_2. G is the compensator operator.

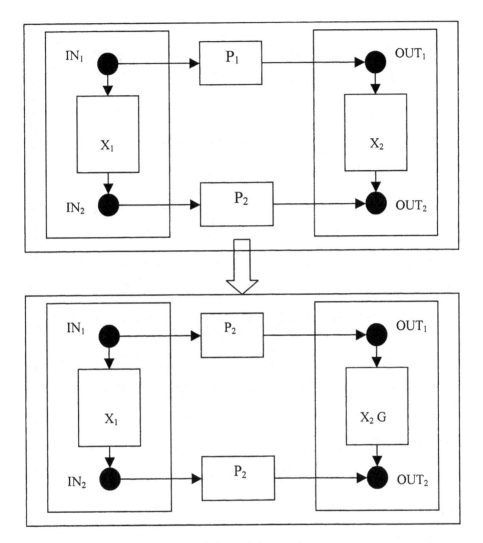

Figure 3.3. The description of the semantic unity which changes one ELS at the second order into another ELS at the second order. Because only an ELS at the third order can change the ELS at the second order, to obtain compensation we must increase the order of the ELS.

The compensator term G can be introduced to change the ELS. We are able to build this change only at the third order of ELS. The same applies when we have the ELS at the second order.

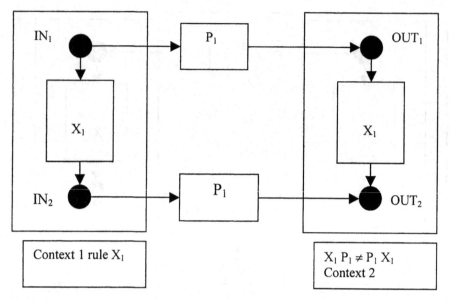

Figure 3.4. A description of the semantic unity or ELS at the second order. Here the coherent equation $X_1 P_1 = P_1 X_1$ is false.

To transform an incoherent ELS as shown in Figure 3.4, we must
introduce a compensator term G. The coherence condition becomes
true when:

$$X_1 G\ P_1\ =\ P_1\ X_1 \tag{3.33}$$

From the (3.33) by means of a simple computation we have:

$$[X_1 G]\ P_1 = P_1\ X_1 + X_1 P_1 - X_1\ P_1$$

or,

$$[X_1 G]\ P_1 = X_1\ P_1 + [P_1\ X_1 - X_1\ P_1]$$

So using the new rule in the second context for which the coherence is valid we obtain:

$$[X_1 G]\ =\ X_1 + [\ P_1\ X_1 - X_1\ P_1\ ]\ P_1^{-1} \tag{3.34}$$

Here we compensate the rule in the second context in such a way as to have a rule similar to X_1 and coherent. This change or compensation can be only obtained when we change one ELS into another in such a way that the compensation is at the third order if the ELS is at the second order.

3.5 Different Models of the Language of GSLT

3.5.1 Affine Transformations and the Covariant Derivative

Consider the GSLT language with the Alphabet $\Sigma .= \{ob_1, ob_2, ob_3, ob_4, op_1, op_2, op_3, op_4, \rightarrow, + \}$ and the variables $V(2) = \{ A^0, A^1, A^2, F^1, F^2, T \}$.

conforming to the usual rules.

From rule R1 or **specification rule,** we obtain:

$$T \Rightarrow A^2$$

$$A^2 \Rightarrow A^1 \rightarrow F^2 \rightarrow A^1 \tag{3.35}$$

Where, $A^1 \rightarrow F^2 \rightarrow A^1$ is a one-dimensional word or ELS(1).

Passing from the one-dimensional word to a two-dimensional word as we can see in (3.3) for $p = 2$ and applying the terminal rule (3.36) we have:

$$A^0 \to F^1 \to A^0 \qquad\qquad ob_1 \to op_1 \to ob_2$$
$$\downarrow \qquad\qquad \downarrow \qquad\qquad\quad \downarrow \qquad\qquad \downarrow$$
$$F^1 \qquad\qquad F^1 \quad\Rightarrow\quad op_2 \qquad\qquad op_4 \qquad\qquad (3.36)$$
$$\downarrow \qquad\qquad \downarrow \qquad\qquad\quad \downarrow \qquad\qquad \downarrow$$
$$A^0 \to F^1 \to A^0 \qquad\qquad ob_3 \to op_3 \to ob_4$$

we obtain:

$$ob_1 \to op_1 \to ob_2$$
$$\downarrow \qquad\qquad\quad \downarrow$$
$$op_2 \qquad\qquad op_4 \qquad\qquad\qquad (3.37)$$
$$\downarrow \qquad\qquad\quad \downarrow$$
$$ob_3 \to op_3 \to ob_4$$

By using the following map:

$$C: \{ob_1 \to V(x)\,, op_1 \to T(x)\,, op_2 \to \partial_x\,, op_3 \to T(x)\,, op_4 \to \partial_x\}$$

$$(3.38)$$

we obtain the model M. In model M, the n-dimensional vector space $V(x) = (v_1(x), ..., v_n(x))$ is considered as an object. $T(x)$ is the affine transformation which depends on the position x of the vector $V(x)$. The operator ∂_x is the following derivative operator ∂_x $= \dfrac{d}{dx}$. By the map C, the model M of A^2 in (3.37) is as follows:

$$V(x) \to T(x) \to T(x)V(x)$$
$$\downarrow \qquad\qquad \downarrow$$
$$\partial_x \qquad\qquad \partial_x \qquad\qquad\qquad (3.39)$$
$$\downarrow \qquad\qquad \downarrow$$
$$\partial_x V(x) \to T(x) \to (T(x) \partial_x V(x), \partial_x (T(x) V(x)))$$

Because the model M is incoherent the final element of ob_4 is given by the two objects $T(x) \partial_x V(x)$ and $\partial_x (T(x) V(x)$. In conclusion we obtain:

$$ob_4 \Rightarrow (T(x) \partial_x V(x), \partial_x (T(x) V(x))) \qquad (3.40)$$

From the model M (3.39) and with the function $\lambda = \lambda_4$ and $\lambda_1 = \lambda_2 = \lambda_3 =$ Identity in (3.31) with simple computation we obtain the model M' in (3.41).

$$V(x) \to T(x) \to T(x)V(x)$$
$$\downarrow \qquad\qquad \downarrow$$
$$\partial_x \qquad\qquad D_x \qquad\qquad\qquad (3.41)$$
$$\downarrow \qquad\qquad \downarrow$$
$$\partial_x V(x) \to T(x) \to (T(x) \partial_x V(x), D_x (T(x) V(x)))$$

where,

$$T(x) \partial_x V(x) = D_x T(x)V(x) \qquad (3.42)$$

Using a mathematical transformation, we can write:

$$D_x T(x)V(x) = \partial_x (T(x) V(x)) + [T(x) \partial_x - \partial_x T(x)] V(x) \qquad (3.43)$$

The compensation operator is connected to the $\partial_x (T(x) V(x))$ term by the operator:

$$[T(x) \, \partial_x - \partial_x \, T(x) \,] \, V(x) \qquad\qquad\qquad (3.44)$$

It has been proved (Mignani et al., 1993) [12] that when $T(x)$ is the **affine transformation** of space-time the operator (3.44) is equal to the geodetic components in the Riemann geometry. The geodetic components are given by the Cristoffel symbols, where D_x is the **covariant derivative** (Mignani et al., 1993). The compensation process is the image by the map C within the compensation element A^3 in (3.30).

$$\begin{array}{l}
V(x) \to T(x) \;\to\; T(x)V(x) \\[6pt]
\qquad\lambda_1 \qquad \partial_x \qquad\qquad \lambda_2 \qquad \partial_x \\[6pt]
V(x) \to T(x) \to T(x)V(x) \\[6pt]
\qquad\qquad\qquad \partial_x V(x) \to T(x) \to (\, T(x)\, \partial_x\, V(x)\,,\, \partial_x\,(\, T(x)\, V(x)\,)\,) \\[6pt]
\partial_x \qquad \lambda_3 \qquad\qquad D_x \qquad\qquad \lambda_4 \\[6pt]
\partial_x V(x) \to T(x) \;\to\; (\, T(x)\, \partial_x\, V(x)\,)
\end{array} \qquad (3.45)$$

where $\lambda_1 = \lambda_2 = \lambda_3$ = Identity operator and

$$\lambda_4 (\, \partial_x\, (\, T(x)\, V(x)\,)\,) = T(x)\, \partial_x\, V(x) = D_x\, T(x)V(x) \qquad (3.46)$$

With the algebraic structure in the model M we can represent the compensation rule. In the (3.46), compensation can also occur by way of transformation of $T(x)$ into $T^*(x)$ as we can see in diagram (3.47).

$$V(x) \rightarrow T(x) \rightarrow T(x)V(x)$$
$$\downarrow \qquad\qquad\qquad \downarrow$$
$$\partial_x \qquad\qquad\qquad \partial_x \qquad\qquad\qquad\qquad (3.47)$$
$$\downarrow \qquad\qquad\qquad \downarrow$$
$$\partial_x V(x) \rightarrow T^*(x) \rightarrow (T^*(x) \, \partial_x V(x) = \partial_x (T(x) \, V(x)))$$

where,

$$T^*(x) \, \partial_x V(x) = T(x) \, \partial_x V(x) + (\partial_x (T(x) \, V(x)) - T(x) \, \partial_x V(x))$$

$$(3.48)$$

The vector $\partial_x V(x)$ changes in the same way as the vector $V(x)$ plus a term of compensation:

$$(\partial_x (T(x) \, V(x)) - T(x) \, \partial_x V(x)) = \frac{dT(x)}{dx} \, V(x) \qquad (3.49)$$

that is, the drag velocity. For complex systems such as Robots (Tzvetkova and Resconi, 1996) [24], we can calculate the drag velocity for general changes of references by using the GSLT.

GSLT was recently used to model affine transformations in non conservative general relativity (Mignani et al.1993) [12].

3.6 GSLT and Approximate Reasoning

The need to find a general framework in which to consider reasoning system approximation derives from a broad range of sources. That is transformation systems, specification libraries, object identification, reasoning re-use and pattern recognition. A categorical approach captures the structural framework well but certain subtleties of approximation are lost when the input-output behaviour common to systems theory is considered. To capture such behaviour we show that General System Logical Theory (GSLT) brings a number of subtle insights to the approximate reasoning (Rattray et al., 1997) [18]. Suppose we have the situation where S_C is an available component specification (algebraic or sketch specification), R_C is a corresponding available realisation or implementation, and A is the known transformation process for developing R_C from S_C. Thus, S_C is the input for A and R_C is the output. Further, suppose that we have a desired component specification S_D which "closely matches" (represented as $k_1: S_C \rightarrow S_D$) the available component specification S_C. What can be said about the degree of realisation which may be associated with S_D? Essentially, our "object of interest" is the transformation B from S_D to R_D.

The intrinsic equation governing the ideal situation ("idealised world") is:

$$B \, o \, k_1 = k_2 \, o \, A \,, \tag{3.50}$$

for some $k_2 : R_C \rightarrow R_D$. In the "real world", this would become

$$B \, o \, k_1 \approx k_2 \, o \, A \tag{3.51}$$

That is the left hand side is only approximately equal to the right hand side. We can analyse the two cases:

$$B \, o \, k_1 \, (S_C) \ \text{and} \ k_2 \, o \, A \, (S_C) \tag{3.52}$$

In terms of their corresponding *intrinsic equations* these represent two different "idealised world" problems (illustrated by the A^2 behind (3.54) and the A^2 in front (3.54)). What we want to identify is the relationship between the two ELS(2) behind and in front (3.54). To capture the essence of the real world situation we have to move to a higher A^3 elementary logical system in (3.54).

The function λ "measures" how far we are from the ideal world and it represents a *compensatory function acting* between the "real world" and the "idealised world".

The compensatory function aims to neutralise the difference between the *true solution in an all-knowing world and the actual solution in the real partial-knowledge world.*

In this example we begin with a **non-commutative** element in (3.53).

$$
\begin{array}{ccc}
S_C \rightarrow k_1 \rightarrow & S_D & \\
\downarrow & \downarrow & \\
A & B & \\
\downarrow & \downarrow & \\
R_C \rightarrow k_2 \rightarrow & (\, B \circ k_1 \,(S_C\,), \; k_2 \circ A \,(S_C\,)\,) &
\end{array}
\tag{3.53}
$$

and with the compensation ELS(3) in (3.54).

$$
\tag{3.54}
$$

where $\lambda_1 = \lambda_2 = \lambda_3 =$ Identity operator .We obtain the compensated ELS(2):

$$
\begin{array}{ccc}
S_C \rightarrow k_1 & \rightarrow & S_D \\
\downarrow & & \downarrow \\
A & & B^* \\
\downarrow & & \downarrow \\
R_C \rightarrow k_2 & \rightarrow & (k_2 \, o \, A \, (S_C))
\end{array}
\qquad (3.55)
$$

and,

$$
\lambda_4 (B \, o \, k_1 \, (S_C)) = k_2 \, o \, A \, (S_C) = B^* \, o \, k_1 \, (S_C) \qquad (3.56)
$$

The new ELS(2) in (3.55) with the compensated transformation B*, is a **commutative** diagram. Unlike the vector space example, this model lacks a well-defined algebraic structure and we cannot define the compensation rule explicitly. The compensation derives from the management process and development tools put in place to ensure the coherence of (3.55).

3.7 GSLT and Active Control of a Field of Waves by Compensation

We have presented the same subject in Chapter 2 as an example of action of the second order. Here we present the same example but from the point of view of the GSLT and so at the abstract level of action representation.

In 1973 Jessel wrote a seminal work on the active control of sound. We want to show that the work of Jessel introduced the compensation rule. We now extend this to different domains of science and technology.

Using the conservation of the mass (continuous equation) and the linear form of the conservation of the momentum (Euler equation), Jessel suggested the system of the differential equations as:

$$c^{-2} \partial_t \, p + \rho_0 \, \text{div } \mathbf{v} = Q$$

$$\rho_0 \partial_t \, \mathbf{v} + \text{grad } p \; = \mathbf{F} \qquad\qquad (3.57)$$

That is, the first order linear equation for a uniform medium at rest (see a perfect fluid, Jessel (page 31, (1973))). Using (3.57) it is easy to calculate the equation of the sound waves as:

$$c^{-2} \partial_{tt} \, p - \text{div grad } p = \partial_t \, Q - \text{div } \mathbf{F} \qquad\qquad (3.58)$$

Here $p(x,y,z,t)$ is the pressure of the medium, $\mathbf{v}$ is the velocity of the displacement of the medium, ρ_0 is the constant density of the medium, c is the velocity of the waves, Q is the monopole density sound source and $\mathbf{F}$ is the bipole sound density source. For simplicity, we write the equation (3.58) symbolically as:

$$\text{OP } \Phi = \mathbf{S} \qquad\qquad (3.59)$$

Where, $\Phi = (p, \mathbf{v})$ is the field of the waves, $\mathbf{S} = (Q, \mathbf{F})$ are the sources. We note that equation (3.59) takes the field as input and obtains the source as output, whereas in physics, $\mathbf{S}$ is the input, or the cause of the output field. With a A [1], the relation between field and source can be given by the ELS(1) as follows:

$$\Phi \rightarrow \text{OP} \rightarrow \mathbf{S} \qquad\qquad (3.60)$$

At this point in the work of Jessel we discovered a new way to change the field Φ. Using a function $T(x,y,z,t)$ the field Φ may be reshaped into another field Φ'. Thus, we obtain another ELS(1) as shown in (3.61).

$$\Phi \rightarrow T \rightarrow \Phi' \tag{3.61}$$

By studying the two simultaneous transformations (3.60) and (3.61) we obtain the ELS(2):

$$\begin{array}{ccc}
\Phi \rightarrow & OP \rightarrow & S \\
\downarrow & & \downarrow \\
T & & T \\
\downarrow & & \downarrow \\
T\Phi \rightarrow & OP \rightarrow & (OP\ T\ \Phi,\ s\ OP\ \Phi)
\end{array} \tag{3.62}$$

where the sources **S** are transformed as follows:

$$T\ S = (T\ Q,\ T\ \Phi) \tag{3.63}$$

Where, T is the function which transforms the field Φ as follows:

$$T\ \Phi = (T\ p,\ T\ v) \tag{3.64}$$

Since Jessel found that the sources T **S** cannot generate the field T Φ, in the language of the GSLT the A^2 in (3.62) does not commute. With the compensation rule we can change the previous A^2 into another A^2 by this ELS(3):

$$\Phi \to OP \to S$$

$$\Phi \to OP \to S$$

$$T\Phi \to OP \to (OP\ T\ \Phi,\ T\ OP\ \Phi)$$

$$T\Phi \to OP \to (OP\ T\ \Phi)$$

(with λ_1, λ_2, λ_3, λ_4, T, T^* mappings in the diagram) (3.65)

where, $\lambda_1 = \lambda_2 = \lambda_3 =$ Identity operator and

$$\Phi \to OP \to S$$

$$T \qquad T^*$$

$$T\ \Phi \to OP \to (\ OP\ T\ \Phi\ ))$$

(3.66)

with,

$$\lambda_4\ (T\ OP\ \Phi) = OP\ T\ \Phi\ =\ T^*\ OP\ \Phi \tag{3.67}$$

With a mathematical transformation, we can write:

$$\lambda_4\ (T\ OP\ \Phi) = T\ OP\ \Phi\ +\ [\ OP\ T - T\ OP\ ]\ \Phi\ =\ T^*\ OP\ \Phi \tag{3.68}$$

Since $S = OP\ \Phi$, the compensation term $[\ OP\ T - T\ OP\ ]\ \Phi\ = S_C$ is again a source that Jessel denoted as secondary sources. The S_C are the sources used to compensate $T\ S$ in such a way as to obtain the field $T\ \Phi$. If the diagram (3.66) does not commute we lose the physical meaning that relates the source with the field. Using a simple computation we can show that:

$S_C = [\text{OP T - T OP}] \Phi = [c^{-2} p \partial_t T + \rho_0 \mathbf{v} \text{ grad } T, \rho_0 \mathbf{v} \partial_t T + p \text{ grad } T]$ (3.69)

Where, $S_C = (Q^s , \mathbf{F}^s)$ are the secondary sources.

In this work, Jessel found the transformation T for which the compensation was not necessary, that is, diagram (3.66) commutes.

In this case $T^* = T$ and:

$[\text{OP T - T OP}] \Phi = [c^{-2} p \partial_t T + \rho_0 \mathbf{v} \text{ grad } T = 0 \quad \rho_0 \mathbf{v} \partial_t T + p \text{ grad } T = 0]$ (3.70)

We can eliminate ρ_0 by multiplying the first equation by $\partial_t s$ and the second equation by -grad s. In this way we obtain the condition for which (3.66) commutes:

$$c^{-2}(\partial_t T)^2 - (\text{grad } T)^2 = 0$$ (3.71)

The transformation T in this case generates a surface that moves in time, perpendicular to the speed c of the wave. This surface is the characteristic surface of the wave equation.

Natural movement of the waves generates a surface of propagation perpendicular to the velocity of the wave. This natural surface T(x,y,z,t) generates the commutative diagram in (3.66) where the field and the source change in the same way.

$$\Phi \rightarrow T \Phi \text{ and } S \rightarrow T S$$ (3.72)

3.8 GSLT in General System Theory

3.8.1 Definitions

Given the system (Wymore, 1993) [25]:

$$Z = < IZ, SZ, OZ, NZ, RZ > \tag{3.73}$$

where IZ is the input set , SZ is the state set , OZ is the output set, NZ : (IZ , SZ) $\to$ SZ is the next state function and RZ : SZ $\to$ OZ is the reply function. Wymore gave the name system Z to this arrangement. The next state function NZ and the reply function RZ can be represented by the language of the GSLT as:

$$A^1 \Rightarrow SZ(t) \to NZ\,(IZ) \to SZ(t+1) \tag{3.74}$$

And,

$$A^1 \Rightarrow SZ(t) \to RZ \to OZ(t) \tag{3.75}$$

which are models of the symbolic form:

$$A^0 \to F \to A^0$$

With the generation rule R3 we use the two generators (3.74) and (3.75) to obtain the ELS(2):

$$
\begin{array}{ccc}
SZ(t) & \xrightarrow{\hspace{1cm}} RZ \xrightarrow{\hspace{1cm}} & OZ(t) \\
\downarrow & & \downarrow \\
NZ(IZ) & & E \\
\downarrow & & \downarrow \\
SZ(t+1) & \xrightarrow{\hspace{1cm}} RZ \to & [\, RZ\,NZ(IZ)\,SZ(t)\,,\,E\,RZ\,SZ(t)\,] = A^2
\end{array}
\tag{3.76}
$$

We note that with two A ^{1} it is possible to define the initial variable SZ(t) in the A ^{2} and the two functions RZ and NZ(IZ). Because RZ is always the same at any time in A ^{2}, we introduce in (3.76) the ELS(1):

$$SZ(t+1) \rightarrow RZ \rightarrow OZ \ (t+1) \tag{3.77}$$

To close A ^{2} we use ELS(1)

$$OZ(t) \rightarrow E \rightarrow OZ(t+1) \tag{3.78}$$

The expectation operator E gives the trajectory in time of the output OZ. In the A ^{2} of (3.76) we have two possible outputs at the same time t+1.

The first output is:

$$RZ \ .NZ(IZ) \ .SZ(t) = OZ1(t+1) \tag{3.79}$$

The second output is:

$$E \ .RZ \ .SZ(t) \ = OZ2(t+1) \tag{3.80}$$

The first output is obtained by using the system function NZ (IZ), the second is given by the expectation operator without using the next state function. When,

$$OZ1(t+1) = OZ2(t+1) \tag{3.81}$$

the system function NZ(IZ) gives as output OZ1(t+1), the expected value OZ2(t+1) generated by the expected operator E. The equation (3.81) is the coherence condition for A ^{2} (ELS(2)) in (3.76). When the equation (3.81) is not true, then A ^{2} in (3.76) is incoherent and it is necessary to change the operators E, NZ(IZ) or RZ in such a way as to have a compensation process and to obtain a new coherent A^2. For (3.81) the formula that relates E, NZ(IZ) and RZ is:

$$E .RZ .SZ(t) = RZ .NZ(IZ) .SZ(t) \tag{3.82}$$

or

$$E\ OZ(t) = NZ(IZ)\ OZ(t) + (\ RZ\ NZ(IZ) - NZ(IZ)\ RZ\ )\ SZ(t)$$

$$\tag{3.83}$$

3.9 Homomorphisms between Systems and GSLT

In the book by Wymore (Wymore, 1993) [25] we found this definition of homomorphism:

"From an engineering point of view, the concept of homomorphism will be employed as the formal definition of when one design has the *functional capability* of another design."

We obtain the homomorphism between systems by the generators:

$$A^1 \Rightarrow SZ1(t) \rightarrow NZ1 \rightarrow SZ1(t+1) \tag{3.84}$$

and

$$A^1 \Rightarrow SZ1(t) \rightarrow H \rightarrow SZ2(t) \tag{3.85}$$

In the two ELS(1), the input IZ1 and IZ2 are constant. H is a function that relates the states of the system Z1 at the time t and the states of Z2 at the time t. With rule R3, we generate the ELS(2).

$$
\begin{array}{ccccc}
\text{SZ1(t)} & \longrightarrow & \text{H} & \longrightarrow & \text{SZ2(t)} \\
\downarrow & & & & \downarrow \\
\text{NZ1} & & & & \text{NZ2} \\
\downarrow & & & & \downarrow \\
\text{SZ1(t+1)} & \longrightarrow & \text{H} & \longrightarrow & \text{SZ2(t+1)}
\end{array}
\qquad (3.86)
$$

When A ^{2} is coherent the design of Z1 has the functional capability of the design of Z2. Using the function H, we can simulate the system Z2 by the system Z1 at time t. The state SZ1(t) is the initial element. Using H and NZ1, we can then calculate the next state of Z1 and the two states of Z2 one at the time t that is SZ2(t) and the other at the time t+1 that is SZ2(t+1). When the ELS(2) is coherent the state SZ1(0) is the free initial state and all the other states for Z1 and Z2 at other times are dependent from the initial state of Z1. For the coherence the state SZ2(t+1) computed by NZ2 is the same as the state computed by H In this case there exists a *homomorphism* between systems Z1 and Z2. When the A ^{2} is incoherent we talk of a *quasi-homomorphism*.

With the GSLT language we can obtain new types of homomorphisms by the generation rule R3 from elementary homomorphisms. *Composing homomorphisms* is a new possibility given by the GSLT. In fact, from the homomorphisms:

$$
\begin{array}{ccccc}
\text{SZ1(t)} & \longrightarrow & \text{H1} & \longrightarrow & \text{SZ2(t)} \\
\downarrow & & & & \downarrow \\
\text{NZ1} & & & & \text{NZ2} \\
\downarrow & & & & \downarrow \\
\text{SZ1(t+1)} & \longrightarrow & \text{H1} & \longrightarrow & \text{SZ2(t+1)}
\end{array}
\qquad (3.87)
$$

and,

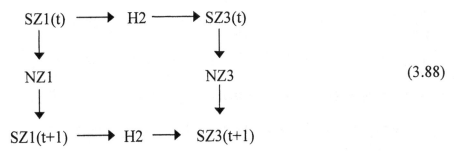

$$(3.88)$$

With rule R3 we can compose (3.87) and (3.88) which give A^3 (ELS(3)):

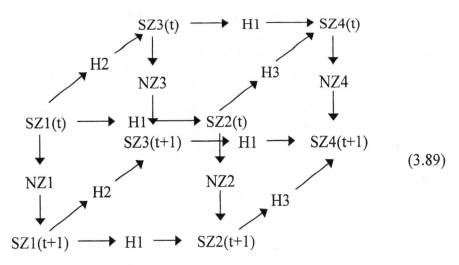

$$(3.89)$$

In A^3 the upper part of the cube is the ELS(2) in (3.90).

$$
\begin{array}{ccccc}
SZ1(t) & \longrightarrow & H1 & \longrightarrow & SZ2(t) \\
\downarrow & & & & \downarrow \\
H2 & & & & H3 \\
\downarrow & & & & \downarrow \\
SZ3(t) & \longrightarrow & H1 & \longrightarrow & SZ4(t)
\end{array}
\qquad (3.90)
$$

where the function H3 is a new type of function given by the co-herence formula:

$$
H1\ H2 = H3\ H1 \quad \text{or} \quad H3 = H1\ H2\ H1^{-1}
\qquad (3.91)
$$

The function H3 is generated by the two elementary functions H1 and H2 by the complex composition. Homomorphisms between Z2 and Z4 by H3 are a new type of homomorphism given by the generating rule of H1 and H2. With rule R3, from Z1 we can simulate Z2 and Z3 by H1 and H2, but we can also simulate Z4 from Z2 and $H3 = H1\ H2\ H1^{-1}$. The function H3 is homomorphic to H2, but H3 is also the homomorphic function from Z2 to Z4. In A^3, from Z1 with H1, H2, H3 we can simulate the systems Z2, Z3, Z4.

Bibliography

1. Beth T., *Keeping secrets a personal matter or the exponential security system*. In Cryptology and Coding III, edited by M. J. Ganley, Oxford University Press, 1993.

2. Fatmi H.A., Marcher P. J., Jessel M. and Resconi G., "Theory of cybernetic and intelligent machine based on Lie commutators." *International Journal of General Systems,* vol. 16, no.2, pp. 123-164, 1990.

3. Higgins P.J., "Algebras with a scheme of operators." *Mathematische Nachrichten*, vol. 27, pp. 115-132, 1963/64.

4. Hill G., *"Category theory for the configuration of complex systems."* In: Proceedings of Third International Conference on Algebraic Methodology and Reasoning Technology, AMAST'93, edited by M. Nivat, C. Rattray, T. Rus and G. Scollo, Workshops in Computing series, Springer-Verlag, London, pp. 193-200, 1994.

5. Jessel M., *Acoustique Theorique (propagation et holophonie).* Masson et Cie Editeurs, Paris, 1973.

6. Kazakov G.A. and Resconi G., "Influenced Markovian checking process by general system logical theory." *International Journal of General Systems,* vol. 22, pp. 277-296, 1994.

7. Klir G.J., *Architecture of Systems Problem Solving.* Plenum Press, New York and London, 1985.

8. Mac Lane S., *Categories for the Working Mathematician, Graduate Texts in Mathematics,* 5, Springer-Verlag, New York, 1971.

9. Manna Z., *Mathematical Theory of Computation*. McGraw Hill, New York, 1974.

10. Mesarovic M.D. and Takahara Y., *Foundations of General System Theory*, Academic Press, 1975.

11. Mesarovic M.D. and Takahara Y., Abstract System Theory, Lecture Notes in Control and Information *Science*, vol.116, Springer-Verlag, 1989.

12. Mignani R., Pessa E. and Resconi G., "Commutative diagrams and tensor calculus in Riemann spaces." *Il Nuovo Cimento*, vol. 108B, no.12, 1993.

13. Mignani R., Pessa E. and Resconi G., "Non-conservative gravitation equations." *General Relativity and Gravitation*, vol.29, no.8, pp.1049 – 1073, 1997.

14. Penna M.P., Pessa E. and Resconi G., "*General system logical theory and its role in cognitive psychology*." Third European Congress on Systems Science, Rome, 1-4 October, 1996.

15. Petrov A.A., Resconi G., Faglia R. and Magnani P.L., "*General system logical theory and its application to task description for intelligent robot*." In: Proceedings of Sixth International Conference on Artificial Intelligence and Information Control of Robots, Smolenize Castle, Slovakia, September 1994.

16. Rattray C., "*Structure algebrique, compilation et programmes structure,* 1iere partie: sigma-schema, sigma-modeles et type de donnes." Research Report No RR4, ENSIMAG, Grenoble, France, 1975.

17. Rattray C. and Rus T., "*Structure algebrique, compilation et programmes structure*, 2ieme partie: vers une technologie de

construction de compilateurs ayant une base theorique." Research Report No RR40, ENSIMAG, Grenoble, France, 1976.

18. Rattray C., Resconi G. and Hill G., "GSLT and reasoning development process", Eleventh International Conference on Mathematical and Computer Modelling & Scientific Computing, Georgetown University, Washington DC, March 31 - April 3 1997.

19. Resconi G. and Jessel M., "A general system logical theory." *International Journal of General Systems*, vol.12, pp. 159-182, 1986.

20. Resconi G. and Tzvetkova G.V., "Simulation of dynamic behaviour of robot manipulators by general system logical theory", *Proceedings of Fourteenth International Symposium 'Manufacturing and Robots'*, Lugano, Switzerland, 25-27 June, pp. 103-106, 1991.

21. Resconi G. and Hill G., "*The language of general system logical theory: a categorical view*." European Congress on Systems Science, Rome, 1-4 October 1996.

22. Rus T., "Algebraic construction of compilers." *Theoretical Computing Science,* vol. 90, pp. 271-308, 1991.

23. Rus T. and Halverson T., "Algebraic tools for language processing." *Computer Languages,* vol.20, no.4, pp. 213-238, 1994.

24. Tzvetkova G.V. and Resconi G., "*Network recursive structure of robot dynamics based on GSLT*." European Congress on Systems Science, Rome, 1-4 October 1996.

25. Wymore A.W., *Model-Based Systems Engineering*. CRC Press 1993.

Chapter Four

Adaptive Agents and Complex Systems

4.1 Introduction

In the definition given by G. Klir, a system S is defined as S = [M, R] where M is a set of objects and R is a set of relations between the objects. We can use this system definition to describe the Logic where the objects are propositions and the relations are the inferential rules. We can also describe the semantics of a linguistic text (semantic web) and many other disciplines. In spite of the success of the system model, we argue that a higher order of systems exists. The systems of order one are ordinary systems with objects and relations. The systems of order two are meta-systems where the relations are transformed into one another. Every transformation changes the relations which are in one context into the image of the same relations in another context. Transformations of relations or functions can be homomorphisms.

When the knowledge is described using a set of different contexts with proper rules, every context C receives images of relations located in other contexts. Within the context C, when the image rules, which come from other contexts, are different from the local rules in C, we have conflict between the two types of rules. To eliminate this conflict, we activate compensation processes. The compensation processes change, at the abstract order, the image relations or the internal relation in C in such a way as to have the same relation R for the images and for the internal relations. Systems of the order three or higher orders are possible.

One of the problems in modern science is posed by the study of complex systems. As it is difficult to define exactly a complex system, we suggest that the classical definition should be enlarged by higher orders of systems as Adaptive Agents. As it is well known, there are many different ways to formalise the concept of complexity. Here we stress that, whatever definition of complexity we adopt, it can always be described by different orders of complexity which are the measure of our difficulty in processing information by only one order of complexity. The difficulty can be due to a number of different reasons:

(1) The information is unavailable which happens in uncertain situations which are the image of new orders of complexity (complexity in uncertain systems).
(2) The information is complete but is available only in an implicit form. To make it explicit, a compensation process in Adaptive Agents is necessary.
(3) The information is complete but is available only in an implicit form. On principle a compensation process at different orders is not always possible in Adaptive Agents or systems.

Many physical systems are in the first case. Stars, Galaxies, the Universe as a whole, systems at a microscopic order and quantum phenomena in the world of atoms and of elementary particles are all possible examples. Considering the second and third case, we remember that they are common in the domain of mathematical models of dynamical systems. In most situations the form of the mathematical equations describing their evolution is known, but on principle, we are unable to solve them, not only from an analytical point of view but also in a number of cases from a numerical point of view.

When we look in a deeper way at these three points we note that we can assemble all the difficulties into one because any information of one object in one context is in conflict with other information of the same object that comes from other contexts. In chapters two and three we denoted this type of conflict as incoherence among the rules in the contexts. The agents that connect contexts as agents of the second order can generate compensation processes to reduce the incoherence.

The conflicts are the origin of the incomplete knowledge and are also the origin of the impossibility of transforming the implicit information into the explicit information. In many cases in special contexts for some agents there is sufficient information but it is not possible to communicate or transport this information to other contexts. In most situations we know the form of the mathematical equations describing their evolution, but we are, on principle, unable to solve them. This is not only in an analytical sense but may be also in a numerical sense. Here when the agent operates in the symbolic context it has all the necessary information. The agent in the symbolic domain may encounter a conflict situation when moving to solutions in a functional or numerical representation.

The agent in its functional or numerical context receives symbolic information from other contexts (a set of differential equations). The symbolic information cannot be reproduced within the numerical context. This generates the well known conflict between symbolic models and functional or numerical representations of the model (set of differential equations and numerical solutions).

The agent in many cases cannot find functions or numerical representations that are coherent when defined in symbolic contexts. This also applies when we have incomplete information on the system being studied. Because a system has different orders of specification the agent also has different contexts to represent the system. At a high order of abstraction the amount of information is limited and in most cases, is complete for the system. When we move to low Adaptive Agent order we are in a situation where the specification of the rules of these Adaptive Agents becomes more complex and more information is necessary. Because we have not always got all the necessary information for any order, the Adaptive Agent is unable to build a coherent set of contexts for all orders of specification. This creates conflict inside the contexts.

Other forms of conflict arise for example from Gödel's theorems, where we can prove a theorem to be true and simultaneously prove it to be false. The theorems for which we cannot decide if there is a proof , and the antinomies show that there are many cases where, given a logical procedure in a domain of symbolic and implicit form, it is not possible to reach a coherent definite context conclusion.

Such a situation requires, when studying complex systems, an attitude which we may call "logical openness". This term is used to denote the fact that, when dealing with complex systems, it must be accepted, as a starting point, that every model is incomplete. The model will be surely falsified by some future experimental findings obtained from the system.

We know that logic creates theories by axioms and theorems. Every logic structure is "closed". If the axioms are true then all the logic consequences derived from the axioms are also true. The complex systems and models contain two different contexts which together create an elementary logical system ELS. The ELS is an open logic object. The agent that uses the ELS operates as an Adaptive Agent at two different orders. The first order relates to the simple observation of the system without using a model. The second order is associated with the adaptation of a previous model to the observation. The previous model is in one context and adaptation is in a new context after obtaining observation.

Because the passage from one context to another is not an inferential process as in logic, in the construction of the model by observation we are not sure that the new model can be coherent with the previous one. The observations are not always in accord with the result of the computation of the previous model. In this case the ELS is logically open and we can have a possible conflict between models and observations. To restore the coherence we change the model by a compensation process. Also when it appears that the model generates results in accord with the observations we are not sure that new observations are not in conflict with the model. This is due to the incompleteness of the observation. A degree of uncertainty or "logical openness" is intrinsic to the ELS compensatory process when a complex system has two or more different models generated using different points of view.

Complex structures grow up for systems or agents of the order three. The structure is made by:

(a) the first context , the system,
(b) the second context, the first model from the first point of view,
(c) the third context , the second model from the second point of view,
(d) the fourth context , the first model from the second point of view or alternatively the second model from the first point of view.

The object we build using these four contexts is an elementary logic system at the third order or ELS(3) or Adaptive Agent of the order three. The ELS(3) is an open logic object with the possibility of many different types of conflicts. The system can generate both observations that are not coherent with the model from the first point of view, and observations that are not coherent with the model from the second point of view. The first model at the second point of view may not be coherent with the second model at the first point of view and so on. To solve all the different types of conflicts in the ELS (3) we must structure the compensation processes in such a way as to avoid any dead lock and to restore the coherence in a synchronic way within all the different contexts. In the Chapters 2 and 3 we show the instruments required to restore the coherence or compensation.

Other objects with "open logic" are possible at a higher order which has a very complex network of conflicts and compensations. In chapter 3 we compared the action of the agents at different orders with the ELS. Here "open logic" means actions at different orders with conflicts and compensations.

We note that any ELS (n) will contain all the other ELS (n-1), ELS(1), ELS (0), where ELS(1) and ELS(0) are unique objects without conflicts. ELS (0) are sets of variables and ELS (1) are sets of operators. All the other objects are "open".

An object ELS (n) is "open" to the input of new information about the operators and variables in the object. Because the new information is not always coherent, the contexts used in ELS (n) will be modified by the operations of compensation until the complete coherence of the object is obtained.

The search for further new information in a well defined direction is triggered by the conflicts between different contexts in the case the coherence cannot be obtained. As it is easy to understand, such a process of mutual interaction between the models and experimental data is very different from an idealised situation codified in the traditional Galilean scientific paradigm. We consider that the process of model modification and context coherence, which occurs in the interaction between model and data, follows well-defined paths. These correspond to prototypical cognitive operations. Such operations may be studied in a rigorous mathematical form. This enables us to design a general scheme for dealing with complex systems in different contexts.

4.2 Representation of Systems of Order One and Two

4.2.1 Systems of Order One

Every system of order one is given by the classical model:

$$S = [M, R] = [M^{(0)}, R^{(1)}] \tag{4.1}$$

Where M are objects and R are relations. Every relation is represented as follows:

$$F = F^{(1)}: A_1 \rightarrow A_2 \tag{4.2}$$

Where $A_1 = F_1^{(0)}$ and $A_2 = F_2^{(0)}$ are sets of objects that belong to M. Sets or objects are considered functions of the order zero. The ordinary relations $F^{(1)}$ are relations of the order one. Every system S in (4.1) in the General System Logical Theory is an ELS(1) or an Elementary Logical System of order one.

4.2.2 Systems of Order Two

Every system of order two is given by the model:

$$S^{(2)} = [M^{(1)}, R^{(2)}] = ELS(2) \tag{4.3}$$

In the system of order two $S^{(2)}$ the relations are given as follows:

$$F^{(2)}: F_1 \rightarrow F_2 \tag{4.4}$$

Where,

$$F_1: A_1 \rightarrow A_2 \quad \text{and} \quad F_2: G_1 A_1 \rightarrow G_2 A_2 \tag{4.5}$$

The relations of order two can also be represented by this diagram:

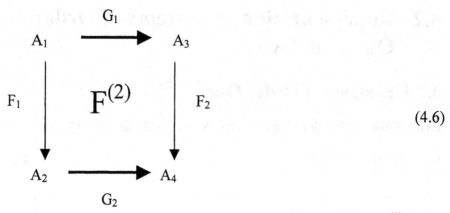

$$(4.6)$$

Where F_2 is an image of F_1. We can write for simplicity, $F_1^{(1)}$ as F_1 and the image $F_2^{(1)}$ as F_2.

4.2.3 Conflict Situations

For the definition of the function F_2 we have:

$$F_2\, G_1\, A_1 = G_2\, F_1\, A_1 \qquad\qquad (4.7)$$

with the constraint,

$$F_2\, G_1 = G_2\, F_1 \qquad\qquad (4.8)$$

All the functions F_2 for which (4.8) is true are projections or images of the function F_1. All the other functions are different from the image. When the function F_2 is not an image of F_1, there is a conflict which we can eliminate by making changes of the functions F_1, G_1, G_2 or at the end of the function F_2.

Note:

Given the function F in a context C_2 and the image of F_1 in C_2, we may say that ELS(2) is incoherent when F is different from the image F_2 of F_1. Any process used to resolve the incoherence or conflict is denoted *compensation*. For the context definition see chapter 2.

The diagram (4.6) can be represented in Figure 4.1 by the contexts C_1 and C_2. In this figure we show the two contexts C_1 and C_2 by two boxes. We put the rule F_1 within C_1 and the rule F_2 within C_2. The external rule F_2 is the image or projection of F_1 into the context C_2. The function F is the internal rule inside C_2 . In figure 4.1 we show F_2 and F in two different boxes .

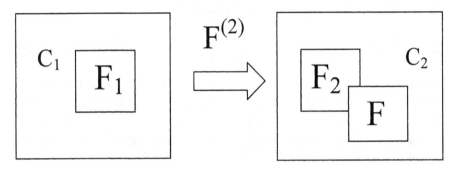

Figure 4.1. In this figure we show the two contexts C_1 and C_2 by two boxes. We show also the function F_1 in C_1, the external function F_2 in C_2 that comes from F_1 and the internal function F that is generated inside C_2. We show by boxes the possible difference between F_2 and F.

In Figure 4.1 by contexts C_1 and C_2 we give an image of the Elementary Logical System at the second order or ELS (2) presented in chapters 2 and 3.

4.2.4 Similarity and Systems

When F_1 is one of the functions of the system $S^{(1)}$ and F_2 is another function of the system $S^{(2)}$ when $G_1 = G_2 = G$ we have F_2: G A_1 $\rightarrow$ G A_2 . In F_2 we have the same structure of F_1 as we proved in chapter 2. When the relation G is "one to one" G is an isomorphism, when G is only "onto" but not "one to one" G is a homeomorphism. Symbolically we use the order relation:

$$F_1 \sqsubset F_2 \tag{4.9}$$

The symbol (4.9) means that there is an order in the similarity between F_1 and F_2. The order means that in $S^{(2)}$ in diagram (4.6) at first we have the function F_1 in the first context and then we project F_1 in the second context in such a way as to obtain F_2.

Example:

Any relation R between the elements belonging to a finite set of objects OB can be represented as follows:

$$R = (OB, K) \tag{4.10}$$

Where K is a set of ordered couples K of OB. We connect any couple with an arrow in a graph and connect any object with one edge in the same graph. So we have a graph representation of the relation R. For example given five objects $\{V_1, V_2, V_3, V_4, V_5\}$ = OB and the set of couples:

$$K = \{(V_5, V_1),(V_1, V_2),(V_1, V_4),(V_2, V_5), (V_3, V_2), (V_3, V_4), (V_5, V_1), (V_5, V_3)\}$$

We have an example of the relation R whose graph is shown in Figure 4.2.

The function F_1 in the diagram (4.6) can be given by the example of relation R previously defined whose graph is shown in Fig. 4.2.

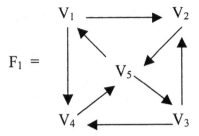

Figure 4.2. In this figure we show the relation F_1 among the five objects $\{V_1, V_2, V_3, V_4, V_5\} = A_1$ (see diagram 4.6 of the second order system) in the context C_1 in Figure 4.1.

Note:

In (4.6), symbolically the function F_1 is represented by only one arrow. But this arrow is the prototype of all possible input output arrows that we can draw in the context C_1. In fact the relation F_1 in Figure 4.2 has eight arrows by which we can obtain the outputs $((V_2, V_4), V_5, (V_2, V_4), V_5, (V_1, V_3))$ from the inputs $\{V_1, V_2, V_3, V_4, V_5\} = A_1$.

For the relations G_1 and G_2 in (4.6) it is possible, for example, to give the relation shown in Figure 4.3.

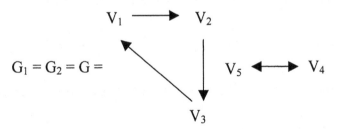

Figure 4.3. In this figure we show the relations $G_1 = G_2 = G$ (see diagram 4.6 of the second order system) between the contexts C_1 and C_2 in Figure 4.1.

With F_1, $G_1 = G_2 = G$, we compute the coherent graph in the context C_2 by the diagram 4.6 of the second order system.

Because F_2: $G\,A_1 \rightarrow G\,A_2$ we obtain the following relation:

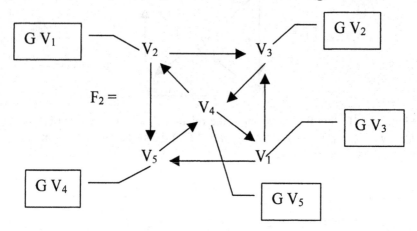

Figure 4.4. In this figure we show the relation F_2 (see diagram 4.6 of the second order system) in the context C_2 in Figure 4.1, obtained by the relation G between the contexts C_1 and C_2.

When we introduce the context C_1 and C_2 the ELS (2) can be represented in this way: we collect all the previous computations in Figure 4.5 where we have the relation F_1 in C_1, the relation G between C_1 and C_2 and the image F_2 of F_1 in the context C_2.

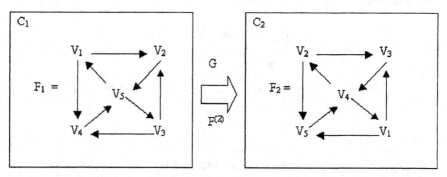

Figure 4.5. In this figure we show the relation F_1 in C_1 and its image F_2 in C_2 (see diagram 4.6 of the second order system) obtained by G.

The $F^{(2)}$ in (4.6) can be represented with full particulars with the Figure 4.6 where the dash arrows are the arrows of the relation G in Figure 4.4.

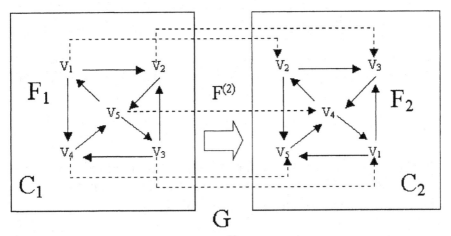

Figure 4.6. Functions $F^{(2)}$ with full particulars.

We note that F_2 is not the same graph as F_1. In fact in F_1 we have derived that we go to V_2 and V_4 from V_1. However in F_2 we go to V_5 and V_3 from V_1. When in the context C_2 we order the vertices as in C_1 we then obtain the relation F_2 shown in Figure 4.7.

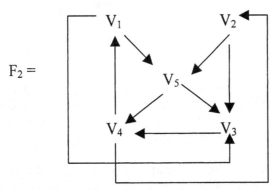

Figure 4.7. Relation F_2 when we order the objects $\{V_1, V_2, V_3, V_4, V_5\}$ in the same way as we order the objects in the relation F_1.

We remark that when we introduce in the second context the same function F_1 that we use in the first place we obtain what is shown in Figure 4.8. When G is the function that joins the first context with the second context, we have the coherence condition $F_2 G = G F_1$, this was shown in chapter 2 to be true.

For example in Figure 4.8 we have $F_1 V_5 = V_3$, $G F_1 V_1 = V_1$, $G V_5 = V_4$ and $F_2 G V_5 = V_5$.. When $F_2 = F_1$, $F_2 G V_5 = F_1 G V_5 = V_5$ and $G F_1 V_5 = V_3$, because V_3 is different from V_5 we have:

$$F_1 G \neq G F_1$$

When the same function is in C_1 and C_2 the relation in C_2 is not coherent with the relation in C_1. So the relation in C_2 is not an image of the relation in C, obtained by means of G.

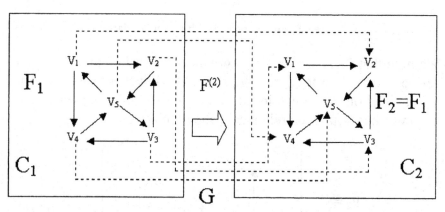

Figure 4.8. In this figure we show the relation F_1 in C_1 and the relation $F_2 = F_1$ in C_2 (see diagram 4.6 of the second order system) obtained by G. We note that $F_1 G V_5 \neq G F_1 V_5$ and the system $F^{(2}$ is not coherent. $F_2 = F_1$ in C_2 is not the image of F_1 in C_1.

When the graph of F_1 in the context C_2 of Figure 4.8 is changed in such a way as to transform this graph into the graph in Figure 4.7, the changes are compensations by which we transform the non coherent system in Figure 4.8 into a coherent system in Figure 4.6. These compensations are well studied in Chapter three.

4.2.5 A Case in Which G is Homomorphism

In chapter 2 we discussed the subject of homomorphism. In this chapter we will give another explanation when we consider functions that belong to systems of the first and second order.

When the transformation G is an onto transformation and is not an one to one transformation we have:

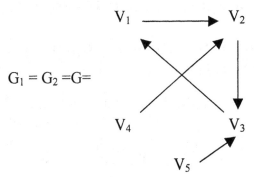

$$G_1 = G_2 = G =$$

Figure 4.9. In this figure we show the new relations $G_1 = G_2 = G$ (see diagram 4.7 of the second order system) between the contexts C_1 and C_2 in Figure 4.1.

In this case we have the function $F^{(2)}$.

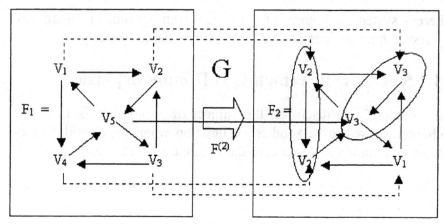

Figure 4.10. Functions $F^{(2)}$ where G is a homomorphism.

4.2.6 Internal Relations to Compensate Non - Similar Relations

In chapter 2 we studied the action of a second order adaptive agent where the transformation of the action of the first order generates a non similar action which we can compensate. In this chapter we discuss the same problem using the complex system theory.

The action of the second order in (4.6) is given by the relation:

$$F_2: G_1 A_1 \rightarrow G_2 A_2 \qquad (4.11)$$

With every couple (A_1, A_2) we can associate a couple $(G_1 A_1, G_2 A_2)$ which is the image of (A_1, A_2).

We choose as example function F_1 in Figure 4.2. For the relations G_1 and G_2 we use as example the graphs in Figure 4.11.

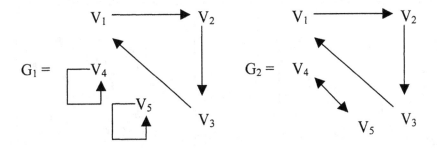

Figure 4.11. We show the relations G_1 and G_2 in (4.6).

The function F_2: $G_1 A_1 \rightarrow G_2 A_2$ is given:

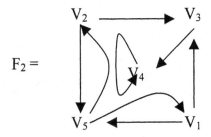

Figure 4.12. We show the relation F_2 in (4.6) obtained by G_1 and G_2 in Figure 4.11 and the relation F_1 in Figure 4.2.

Because F_2 is not similar to F_1 we have:

$$F_1 \not\cong F_2$$

And the function $F^{(2)}$ in Figure 4.13.

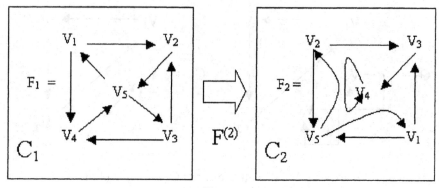

Figure 4.13. Functions $F^{(2)}$ when G_1 and G_2 are different.

Given the coherent equation:

$$F_2 G_1 = G_2 F_1$$

shown in chapter two where $G_1 = P_1$ and $G_2 = P_2$, we are able to compensate F_2 in such a way as to restore the similarity between the two contexts. We have:

$$F_2 G G_2 = G_2 F_1$$

Where $G G_2 = G_1$ and $G = G_1 G_2^{-1}$ is the compensator.

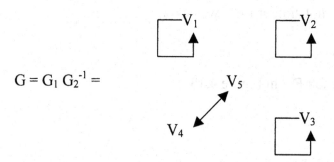

Figure 4.14. Compensator $G = G_1 G_2^{-1}$.

The function F_2 can be compensated by the function $F_2 G$ as shown in Figure 4.15.

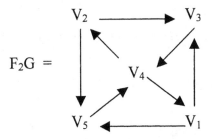

Figure 4.15. Function $F_2 G$ is similar to F_1.

We remark that the $F_2 G$ is not equal to F_1 in Figure 4.13 but has the same structure. So we write:

$$F_1 \sqsubseteq F_2 G$$

When we draw only the arrows in F_1 and substitute the objects with white boxes, the result is the structure of F_1 as shown in Figure 4.16.

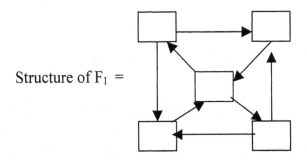

Figure 4.16. Structure of F_1.

The structure of F_1 has input arrows that go into the boxes and output arrows that go from one box to another. When we consider the structure as an INVARIANT and put different objects in the boxes to connect objects in the same way as in function F_2, we have the structure in Figure 4.17.

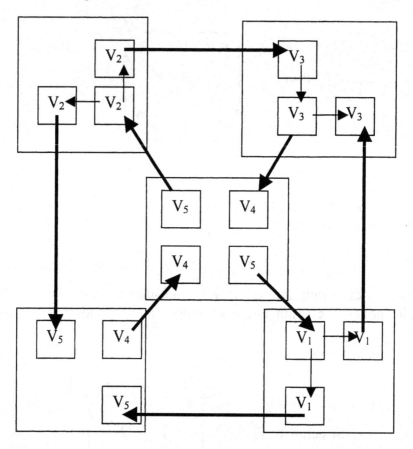

Figure 4.17. In this figure by means of the structure of F_1 we show the function F_2. In the function F_2 from V_1 we go to V_5 and from the same point V_5 comes back to V_1. In this figure from V_1 we go to V_5 in one box and from V_5 in another box we come back to the object V_1. In the same box the input object is not always connected with the output object. The separation of the input from output in the same box is due to the non similarity of F_1 to F_2 (F_1 and F_2 have not the same structure).

We note that in Figure 4.17 we have the same structure as F_1 for the bold arrows but not all the input objects are connected with the

output objects in the same box. When we use the graph G to connect objects inside the boxes, we obtain the relation F_2 G that is
similar to F_1. In Figure 4.18 we show the boxes and the connections.

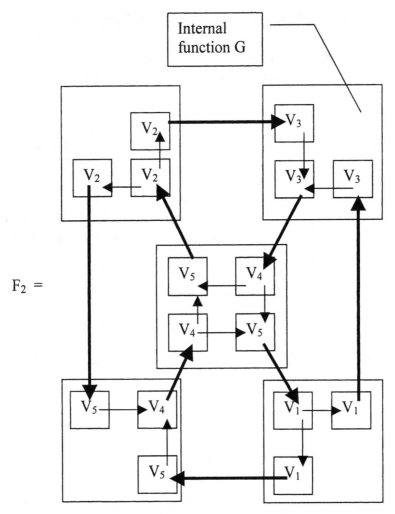

Figure 4.18. Relation F_2 G represented by internal arrows, arrows of G
and external arrows or arrows of F_2. We separate the internal part from
the external part of F_2 G that is similar to F_1.

In Figure 4.18 we show the function F_2 as an external part of the boxes with the same structure as F_1 and the internal part where the arrows of G connect the input and the output objects.

4.3   Examples of Systems of Order Two

A common feature of the model system Klir (1985) [26] consists of a set of objects M and a set of relations R among the objects.

$$S = [M, R] \tag{4.12}$$

where S is the system. For our purpose we write the model as:

$$S^{(1)} = [M^{(0)}, R^{(1)}] \tag{4.13}$$

where $S^{(1)}$ is the system S, $M^{(0)} = M$ are objects, $R^{(1)} = R$ are ordinary relations, We write (1) as apex in order to stress that the ordinary system S is a system of the order one.

For the order two we can define the meta-system as follows:

$$S^{(2)} = [M^{(1)}, R^{(2)}] \tag{4.14}$$

Where $M^{(1)}$ are sets of relations among objects and $R^{(2)}$ are relations of the type two. Let us express:

$$M^{(1)}: OB_1 \rightarrow OB_2 \tag{4.15}$$

And,

$$R^{(2)}: M^{(1)}_1 \rightarrow M^{(1)}_2 \tag{4.16}$$

Where OB_1 is one object of M and OB_2 is another example of M.

Example:

In chapter 2 we gave one simple example of system of the order two as an "adaptive channel "and here we give other examples to clarify the possible applications of the abstract representation of $F^{(2)}$ in Figure 4.1.

Example:

When we separate the procedure from the process in such a way that the procedures are the context and the processes are the functions, we have $S^{(2)}$.

As shown in an abstract way in Figure 4.1, $F^{(2)}$ can be represented as one transformation from one context C_1 to C_2. In a more concrete case we represent the process migration as a transformation between two contexts as we show in Figure 4.19.

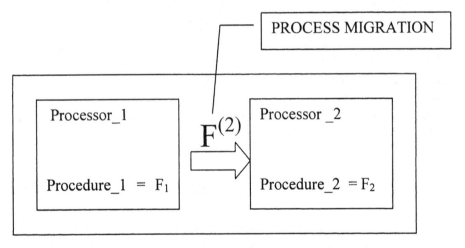

Figure 4.19. The process of migration from procedure_1 = F_1 to procedure_2 = F_2 running in two different processors or computers (contexts C_1 and $C_{2)}$ can be represented as one system at the second order whose abstract structure was represented in Figure 4.1.

The same structure in Figure 4.18 can be also expressed as follows

PROCESS MIGRATION $(F^{(2)})$

Super context = PROCESS MIGRATION (context that includes C_1 and C_2)
Contexts = PROCESSORS or COMPUTERS (contexts C_1 or C_2)
Rule in the context = PROCEDURE (relations or functions F_1, F_2)

Aim of the second order system in Figure 4.18:

A process achieved by running Procedure_1 on Processor_1 will be achieved by running Procedure_2 on Processor_2.

The system of order two in Figure 4.18 is:

$S^{(2)}$ = [PROCEDURE, PROCESS MIGRATION]

Where PROCEDURE = (Procedure_1 = F_1, Procedure_2 = F_2) = $M^{(1)}$ and PROCESS MIGRATION = $R^{(2)}$ = $F^{(2)}$.

Examples similar to 4.19 can be shown in Figures 4.20, 4.21, and 4.22.

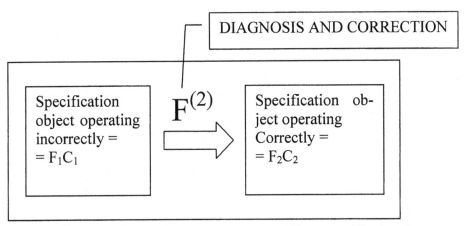

Figure 4.20. Where context C_1 and C_2 are two different specifications for the objects. One specification operates incorrectly the other specification after diagnosis and correction $F^{(2)}$ operates correctly.

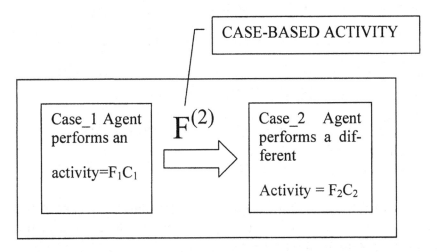

Figure 4.21. Where context C_1 and C_2 are two different cases. In one case the agent performs one activity, in another case the agent performs another activity. The second order transformation $F^{(2)}$ changes one activity in one case into another activity in another case.

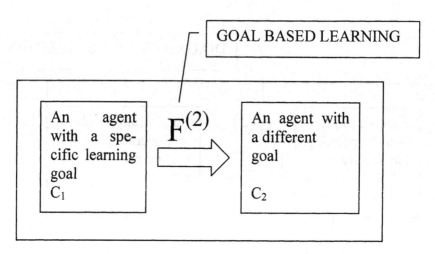

Figure 4.22. Where context C_1 and C_2 are two different goals. In one context the agent learns to achieve one goal by the process F_1, in the other context the agent learns to achieve another goal by the process F_2. The second order transformation $F^{(2)}$ changes one goal into another.

The same structures in Figure 4.20, 4.21 and 4.22 can be also expressed as follows:

FAULT DIAGNOSIS AND CORRECTION

> Super context = DIAGNOSIS AND CORRECTION
> Contexts = SPECIFICATION
> Rules in the contexts = DESIGN (F_1, F_2)

If there are objects operating incorrectly, the same objects need correction to operate correctly (compensation) in their specification context C_2 with suitable design F_2.

$S^{(2)}$ = [SPECIFICATION, DIAGNOSIS AND CORRECTION]

CASE-BASED ACTIVITY

> Super context = CASE-BASED ACTIVITY

Contexts = CASE
Rules in the contexts = ACTIVITY

An agent in Case_1 performs a specific activity; the same agent in Case_2 requires a different activity. Given case_1 and its associated activity and case_2, it is necessary to compensate the activity of the first case to give activity to case_2.

$S^{(2)} = [CASES, ACTIVITY]$

GOAL BASED LEARNING

Super context =GOAL BASED LEARNING
Contexts = GOAL
Rules in the contexts = BODY OF KNOWLEDGE

An agent with a body of knowledge sufficient to achieve a specific learning goal, requires to compensate the body of knowledge in C_1 with a body of knowledge in C_2 to achieve a different learning goal.Given the initial goal and the initial body of knowledge for an agent we compensate the body of knowledge of the agent to give a new goal.

$S^{(2)} = [LEARNING GOAL, GOAL BASED LEARNING]$

4.4 Representation of the ELS(3) or $\mathbf{F^{(3)}}$

In chapters 2 and 3 we have shown the action of the adaptive agent with order three, four and higher. In this chapter we want to present the same extensions of the orders but with the function definition in the systems at different orders. We always use the formal simplification for which the relation of function F is written as $F^{(1)}$. For the function formalism the ELS(3) can be written as follows:

$$F^{(3)}: F^{(2)}_1 \rightarrow F^{(2)}_2 \qquad (4.17)$$

Where,

$$F^{(2)}{}_1: F_1 \rightarrow F_2 \quad , \quad F^{(2)}{}_2: G^{(2)}{}_1 F_1 \rightarrow G^{(2)}{}_2 F_2 \tag{4.18}$$

And,

$$F_1: A_1 \rightarrow A_2 \quad , \quad F_2: G_1 A_1 \rightarrow G_2 A_2 \tag{4.19}$$

$$G^{(2)}{}_1: F_1 \rightarrow F_3 \quad , G^{(2)}{}_2: F_2 \rightarrow F_4 \tag{4.20}$$

With,

$$F_3: G_5 A_1 \rightarrow G_6 A_2 \quad , F_4: G_7 G_1 A_1 \rightarrow G_8 G_2 A_2 \tag{4.21}$$

For,

$$F^{(2)}{}_2: F_3 \rightarrow F_4 \tag{4.22}$$

We have:

$$F_4: G_3 G_5 A_1 \rightarrow G_4 G_6 A_2 \tag{4.23}$$

Because we have two different representations of F_4 we have the identities:

$$G_7 G_1 A_1 = G_3 G_5 A_1 \quad \text{and} \quad G_8 G_2 A_2 = G_4 G_6 A_2 \tag{4.24}$$

For the previous identities we have:

$$G_3: G_5 A_1 \rightarrow G_3 G_5 A_1 \text{ and } G_4: G_6 A_2 \rightarrow G_8 G_2 A_2 \tag{4.25}$$

In function $F^{(3)}$ we have different types of conflicts. For the previous definitions we have:

$$F_1 G_1 A_1 = G_2 A_2 = G_2 F_1 A_1 \tag{4.26}$$

$$F_3\, G_5\, A_1 = G_6\, A_2 = G_6\, F_1\, A_1 \tag{4.27}$$

$$F_4\, G_7\, G_1\, A_1 = G_8\, G_2\, A_2 = G_8\, G_2\, F_1\, A_1 \tag{4.28}$$

$$F_4\, G_3\, G_5\, A_1 = G_4\, G_6\, A_2 = G_4\, G_6\, F_1\, A_1 \tag{4.29}$$

$$G_7\, G_1\, A_1 = G_3\, G_5\, A_1 \quad \text{and} \quad G_8\, G_2\, A_2 = G_4\, G_6\, A_2 \tag{4.30}$$

When all the equations are true $F^{(3)}$ is coherent and every conflict is eliminated.

For the function at the second order $F^{(2)}$, when coherence is obtained and $G_1 = G_2$, we have the similarity order relation in (4.9) . In the $F^{(3)}$ we have four functions F_1, F_2, F_3, F_4 . When coherence is obtained and $G_1 = G_2 = G_3 = G_4$ we have the similarity order relation (Lattice) in Figure 4.23.

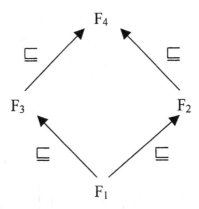

Figure 4.23. Similarity order relation in the coherence case among the functions F_1, F_2, F_3, F_4 when $G_1 = G_2 = G_3 = G_4$.

As shown in (4.6) all that has been previously said on $F^{(3)}$ can be summarized in only one object shown in Figure 4.24 as follows.

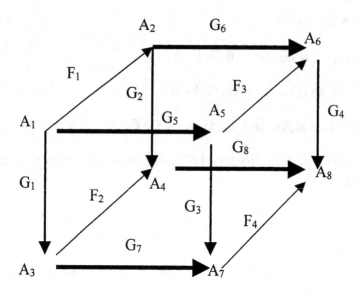

Figure 4.24. Graphic image of the atomic elements for the function at the third order $F^{(3)}$.

The same object given in Figure 4.24 can be shown for simplicity also in a planar image as shown in Figure 4.25.

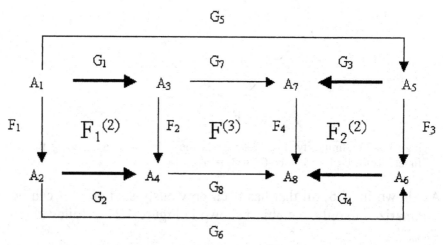

Figure 4.25. Planar image of the object $F^{(3)}$ given in Figure 4.24.

As we have shown in Figure 4.1 for the function of the second order $F^{(2)}$, we can similarly represent the $F^{(3)}$ in Figure 4.26 where we put in evidence the contexts where $F^{(3)}$ operates.

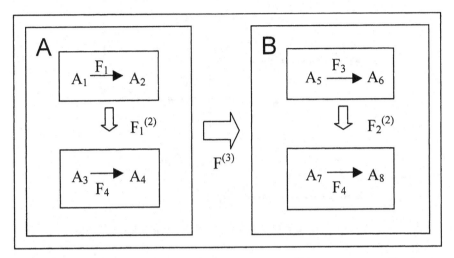

Figure 4.26. Context image of the function $F^{(3)}$ in the system $S^{(3)}$.

Example:

Given the two relations

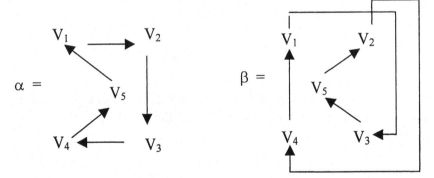

we have the $F^{(3)}$

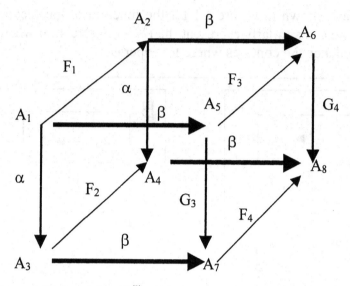

Figure 4.27. Example of $F^{(3)}$ where $G_1 = G_2 = \alpha$ and $G_5 = G_6 = G_7 = G_8$.

When F_1 is the relation:

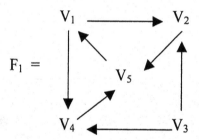

As we have shown in Figure 4.6, a complete example of $F^{(2)}$, similarly we can show a complete example of the function $F^{(3)}$.

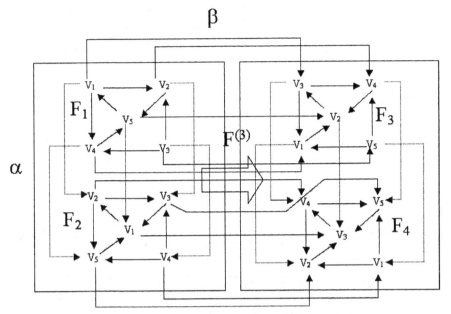

Figure 4.28. Complete example of the $F^{(3)}$ in Figure 4.25.

4.5   Representation of the ELS(4) or $F^{(4)}$

The $F^{(4)}$ can be written as follows:

$$F^{(4)}: F^{(3)}_1 \rightarrow F^{(3)}_2 \tag{4.31}$$

Where,

$$F^{(3)}_1: F^{(2)}_1 \rightarrow F^{(2)}_2, \; F^{(3)}_2 : G^{(3)}_1 \; F^{(2)}_1 \rightarrow G^{(3)}_2 \; F^{(2)}_2 \tag{4.32}$$

And,

$$F^{(2)}_1: F_1 \rightarrow F_2, \; F^{(2)}_2 : G_1^{(2)} \; F_1 \rightarrow G_2^{(2)} \; F_2 \tag{4.33}$$

As in Figure 4.25 we have shown the function at the third order $F^{(3)}$, in Figure 4.29 we show the function $F^{(4)}$.

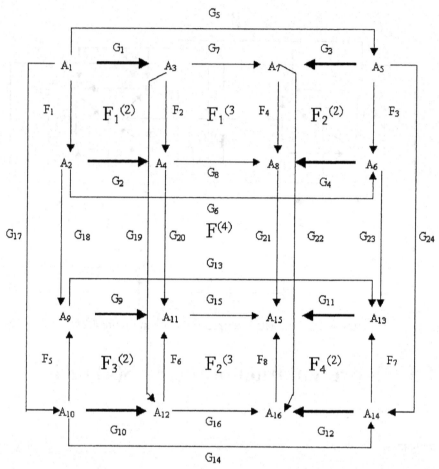

Figure 4.29. Planar image of the transformation at the fourth order $F^{(4)}$.

As we have shown in Figure 4.23 the order of the different functions, we can repeat the same for the function at the fourth order $F^{(4)}$ in Figure 4.30.

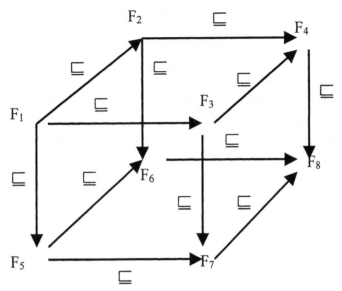

Figure 4.30. Order relation for functions in $F^{(4)}$.

4.6 Conclusion

In this chapter an extension to ordinary definition of relations has been given. Every relation $F^{(n)}$ is of order n and is defined in general by the form:

$$F^{(n)}: F^{(n-1)}_{1} \rightarrow F^{(n-1)}_{2} \tag{4.34}$$

As we have previously shown, the relation $F^{(2)}$ can be written as in (4.11), $F^{(3)}$ can be written as in (4.17), $F^{(4)}$ can be written as in (4.31), so the expression (4.34) is the natural extension of the expressions (4.11), (4.17) and (4.31).

Every order creates a set of conflicts that must be solved. The origin of the conflicts is the difference between the relations inside one context and the relations that come from other contexts. When the two types of relations are the same no conflict exists. When relations that come from external contexts and from internal contexts are different we have a conflict and we activate a compensation process to obtain an identity between all internal and external relations and conflicts can be resolved. The ordinary system is based on the relation definition (ontology) causing us to think that a more abstract order of systems can be built to control the relations at different orders. In computer science we think that the different orders of transformations $F^{(n)}$ may be used to formulate in a mathematical way complex communication among computers.

Bibliography

1. Penna M.P., Pessa E. and Resconi G., General System Logical Theory and its role in cognitive psychology, Third European Congress on Systems Science, Rome, 1-4 October 1996.

2. Tzvetkova G.V. and Resconi G., Network recursive structure of robot dynamics based on GSLT, European Congress on Systems Science, Rome, 1-4 October 1996.

3. Resconi G. and Tzvetkova G.V., Simulation of Dynamic Behaviour of robot manipulators by General System Logical Theory, 14-th International Symposium "Manufacturing and Robots" 25-27 June, pp.103-106.,Lugano Switzeland,1991

4. Resconi G. and Hill G., The Language of General Systems Logical Theory: a Categorical View, European Congress on Systems Science, Rome, 1-4 October 1996.

5. Rattray C., Resconi G. and Hill G., GSLT and software Development Process, Eleventh International Conference on Mathematical and Computer Modelling and Scientific Computing, Georgetown University, Washington D.C.March 31-April 3,1997,

6. Mignani R., Pessa E. and Resconi G., Commutative diagrams and tensor calculus in Riemann spaces, Il Nuovo Cimento, vol.108B, no.12. December 1993.

7. Petrov A.A. Resconi G., Faglia R. and Magnani P.L., General System Logical Theory and its applications to task description for intelligent robot. In Proceeding of the sixth International Conference on Artificial Intelligence and Information Control System of Robots, Smolenize Castle, Slovakia, September 1994.

8. Minati G. and Resconi G., Detecting Meaning, European Congress on Systems Science, Rome, 1-4 October 1996.

9. Kazakov G.A. and Resconi G., Influenced Markovian Checking Processes By General System Logical Theory, International Journal General System, vol.22, pp.277-296, 1994.

10. Saunders Mac Lane, *Categories for Working Mathematician*, Springer-New York, Heidelberg Berlino, 1971.

11. Wymore A.W., *Model-Based Systems Engineering*. CRC Press, 1993.

12. Mesarovich M.D. and Takahara Y., *Foundation of General System Theory*. Academic Press, 1975.

13. Mesarovich M.D. and Takahara Y., Abstract System Theory, *In Lecture Notes in Control and Information Science 116*. Springer Verlag 1989.

14. Resconi G. and Jessel M., A General System Logical Theory, International Journal of General Systems, vol.12, pp.159-182, 1986.

15. Resconi G. and Wymore A.W., Tricotyledon Theory of System amd General System Logical Theory, Eurocast'97, 1997.

16. Fatmi H.A., Marcer P.J., Jessel M. and Resconi G., "Theory of Cybernetics and Intelligent Machine based on Lie Commutators", vol.16, no.2, pp123-164, 1990.

17. Kalman R.E., Falb P.L. and Arbib M.A., *Topics in Mathematical System Theory*. McGraw--Hill Publ., 1969.

18. Mesarovic M.D. and Takahara Y. Abstract Systems Theory, Lecture Notes in Control and Information, 1989 Systems, Springer—Langer, 1989.

19. Padulo L. and Arbib M.A., *System Theory*. WB Saunders, 1974.

20. Resconi G., Rattray C. and Hill G., "The Language of General Systems Logical Theory (GSLT)", *International Journal of General System*, vol. 28, no.4-5, pp.383-416, 1999.

21. Rattray C., "Identification and Recognition through Shape in Complex Systems", 1996.

22. Marshall G.J. and Behrooz A., "Adaptation Channels, Cybernetics and Systems", vol.26, no.3, pp.349-365, 1995.

23. Santilli R.M., Foundation of Theoretical Mechanics, vol I and II: Birkhoffian Generalization of Hamiltonian Mechanics, Springer Verlag, Heidelberg/NewYork, 1982.

24. James A. Crowell, Martin S. Banks, Krishna V. Shenoy and Richard A. Andersen, Visual self-motion perception during head turns, nature neuroscience, vol.1, no.8, pp.732-737, 1998.

25. Array F., Norman R.Z. and Cartwright B., *Introduction à la thèorie des graphes orientés*, Dunod Paris, 1968.

26. Klir G. *Architecture of System Problem Solving*. Prenum Press, New York, London, 1985.

27. Lin Y., "Development of New Theory with Generality to unify diverse disciplines of knowledge and capability of applications", *Int. J. General System*, vol.23, pp.221-239, 1995.

28. Gurevich Y., "Sequential Abstract State Machines Capture Sequential Algorithms", *ACM Transactions on Computational Logic*, vol.1, no.1, pp.77-111, July 2000.

Chapter Five

Adaptive Agents and Models of the Brain Functions

5.1 Introduction

In the previous chapters we have presented the action of the adaptive agent at different orders with its abstract and functional image. Now we give a model of the action that is considered to be similar to the brain functions. We then argue that the brain is a model of the action of the adaptive agent. In an important paper Alexandre Pouger and Lawrence H. Snyder [14] use the basis functions as the main frame to model brain functions. In the abstract of the paper they state:

"Behaviours such as sensing an object and then moving your eyes or hand toward it require that sensory information be used to help generate a motor command, a process known as a sensorimotor transformation. Here we review models of sensorimotor transformations that use a flexible intermediate representation that relies on *basis function*. The use of basis function as an intermediate is borrowed from the theory of non linear function approximation. We show that this approach provides a unifying insight into the neural basis of three crucial aspects of sensorimotor transformations, namely, computation, learning and short term memory. This mathematical formalism is consistent with the responses of cortical neurons and provides a fresh perspective on the issue of frames of reference in spatial representations".

In this abstract we can see that every model of brain functionality must have a basis function to create the context where we locate processes of the brain as the sensory-motor transformations.

In this chapter we go beyond the simple use of the basis function as in the case of the Fourier or Laplace transformation. In conformity with the previous chapters, we generate a context by using a n-dimensional non - Euclidean space of features which are in this particular case the basis functions.

The space of the features is the main reference in the n-dimensional space where we locate our objects and rules. In this space we can define distances and use the vector space properties or rules. The space of the *features* is the prototype of the context used for definition in the previous chapters. This space feature is also used in the Learning Machine, the Support Vector Machine and Learning with a kernel [15].

In this book we use the space of the features or the space of the basis functions in a way different from the learning machine. In this chapter we proceed in a new direction in the use of the space of the features. The direction is in accord with the other chapters of this book where we study the action at different orders of the adaptive agent.

5.2 Description of the Context by Reference to the Features

Given a context with a set of basis features (attributes) ψ_1, ψ_2, ..., ψ_q whose values are represented in Table 5.1.

Table 5.1. Samples of values for the functions in (5.1).

	ψ_1	ψ_2	.	ψ_q
Ob_1	X_{11}	X_{21}	.	$X_{q\,1}$
Ob_2	X_{12}	X_{22}	....	$X_{q\,2}$
..	.		..	......
Ob_{n-1}	X_{1n-1}	$X_{2\,n-1}$	...	$X_{q\,(n-1)}$
Ob_n	X_{1n}	$X_{2\,n}$		$X_{q\,n}$

Table 5.1 connects the n-dimensional space of the objects.

The q vectors in the n-dimensional space are:

$$\psi_1 = (X_{11}, X_{12}, \ldots\ldots, X_{1\,n-1}, X_{1\,n}) \tag{5.1}$$

$$\psi_2 = (X_{21}, X_{22}, \ldots\ldots, X_{2\,n-1}, X_{2\,n})$$

$$\psi_q = (X_{q1}, X_{q2}, \ldots\ldots, X_{q\,n-1}, X_{q\,n})$$

In the same table we have vectors of the features in the q dimensional space:

$$Ob_1 = (X_{11}, X_{21}, \ldots\ldots, X_{q-1\,1}, X_{q\,1}) \tag{5.2}$$

$$Ob_2 = (X_{12}, X_{22}, \ldots\ldots, X_{q-1\,2}, X_{q\,2})$$

$$Ob_n = (X_{1\,n}, X_{2\,n}, \ldots\ldots, X_{q-1\,n}, X_{q\,n})$$

Example 1

Given two colors or features and three objects Table 5.1 becomes Table 5.2:

Table 5.2. Samples of the values for two colors and three objects.

	Red	Green
Table	1	0
Chair	0	1
Lamp	1	1

The vector of the objects are two vectors in a three-dimensional space:

$Red = \psi_1 = (1, 0, 1)$

$Green = \psi_2 = (0, 1, 1)$

The vectors of the features are three vectors each of two dimensions:

$Table = \mathbf{Ob}_1 = (1, 0)$

$Chair = \mathbf{Ob}_2 = (0, 1)$

$Lamo = \mathbf{Ob}_3 = (1, 1)$

5.2.1 Components in the Space Generated by the Feature Vectors in an Object Space

When we assume that the features are the coordinates of our space that define our context, then every vector in this space can be written as a superposition of the coordinates or features as follows:

$$V^k = \sum_{j=1}^{q} w^j X_{j,k} \qquad\qquad (5.3)$$

where w^j are the contro-variant components of the vector V_i where $i = 1, 2, \quad ., n$.

The vector V_i is a vector in the subspace of the objects included in the space generated by the vectors of the feature.

Example 2

The space of the features of the colors for a vector V can be written as:

$$V = w^1 (1, 0, 1) + w^2 (0, 1, 1)$$

When $w^1 = 2$ and $w^2 = 0.5$, we have:

$$V = 2 (1, 0, 1) + 0.5 (0, 1, 1) = (2, 0.5, 2.5)$$

When this is represented in a graphic manner we have Figure 5.1.

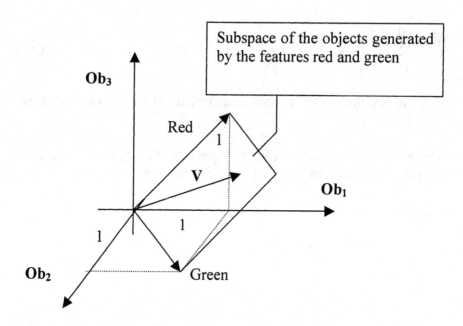

Figure 5.1. The space of the objects and the features. Here the space of the features defines a sub-space of the objects. Here are shown all possible vectors **V** obtained from the weighted superposition of the vectors of the features. The weights of the superposition are the contro-variant components of the vector **V**.

In Figure 5.2 the contro-variant components of the vector V are shown in the space of the features: in general it is not a Euclidean space. The coordinates of the features are general.

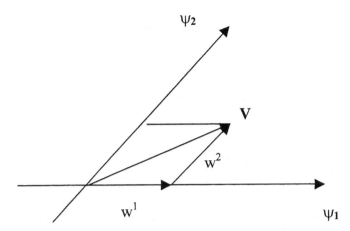

Figure 5.2. Example of the contro-variant components in two-dimensional feature space. In this figure two generic vectors arbitrarily represent the vectors ψ_1, ψ_2 of the feature.

There are two types of general coordinates. The contro-variant components and the covariant coordinates may be defined in this way:

$$w_j = \sum_k V^k X_{j,k} \tag{5.4}$$

In Figure 5.3 an example of the covariant coordinates for two dimensions is shown.

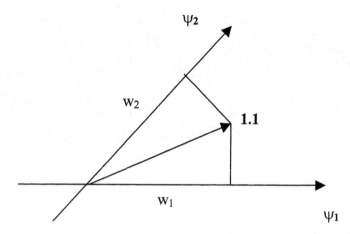

Figure 5.3. Example of the covariant components in two dimensional feature space.

When equation (5.3) is substituted in (5.4) we obtain:

$$w_j = \sum_i \sum_k w^i X_i^k X_{j,k} = \sum_i w^i g_{i,j} \qquad (5.5)$$

where,

$$g_{i,j} = \sum_k X_i^k X_{j,k}$$

This gives the relationship between the two types of components in the feature space.

In Figure 5.4 we show the orthonormal or Euclidean case where $g_{i,j}$ is a unity matrix. In the orthonormal system or Euclidean space all the covariants and controvariant components of a vector are equal.

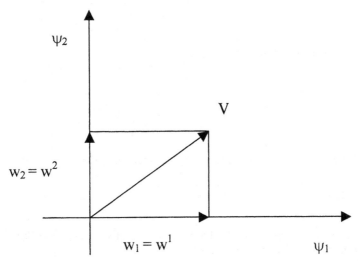

Figure 5.4. The Euclidean coordinates where the contro-variant and the covariant components are equal.

Remark:

It is noted that using general coordinates the distance between two points is given by the equation:

$$S = \Sigma\ g_{i,j}\ w^i\ w^j \qquad\qquad (5.6)$$

When we have Euclidean coordinates the distance S is the ordinary distance given by:

$$S = (w^1)^2 + (w^2)^2 + \qquad .(w^q)^2 \qquad\qquad (5.7)$$

Example 3

For a Euclidean space where:

$$g_{i,j} = \begin{matrix} 1 & 0 \\ 0 & 1 \end{matrix} \qquad (5.8)$$

The distance between two points is given by the equation:

$$S = x_1^2 + x_2^2 \qquad (5.9)$$

In Figure 5.5 we show in the Euclidean space the points having the same distance from the centre.

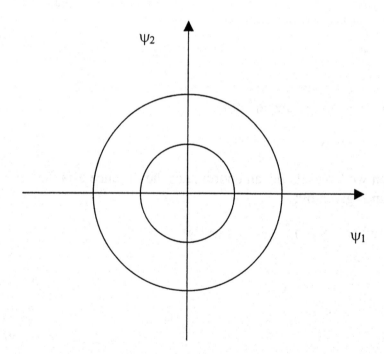

Figure 5.5. Points at the same distance in the Euclidean space when the distance is given by $S = x_1^2 + x_2^2$.

When we have:

$$g_{i,j} = \begin{array}{cc} 1 & \cos(\theta) \\ \cos(\theta) & 1 \end{array} \qquad (5.10)$$

the expression of the distance is:

$$S = x_1^2 + x_2^2 + 2\cos(\theta)\, x_1\, x_2 \qquad (5.11)$$

In Figure 5.6 we show the ellipse which represents the line at the same distance in the general coordinates:

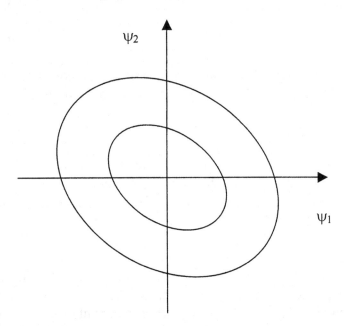

Figure 5.6. Points at the same distance in the general coordinate space or non Euclidean space, for a particular angle. The distance is $S = x_1^2 + x_2^2 + 2\cos(\theta)\, x_1\, x_2$.

The general expression for the tensor $g_{i,j}$ is given by the matrix:

$$g_{i,j} = \begin{bmatrix} \sum\limits_{i=1}^{n} X_{1,i}^{2} & \sum\limits_{i=1}^{n} X_{1,i} X_{2,i} & \cdot\cdot & \sum\limits_{i=1}^{n} X_{1,i} X_{(n-1),i} & \sum\limits_{i=1}^{n} X_{1,i} X_{n,i} \\[2ex] \sum\limits_{i=1}^{n} X_{1,i} X_{2,i} & \sum\limits_{i=1}^{n} X_{2,i}^{2} & \cdot\cdot & \sum\limits_{i=1}^{n} X_{2,i} X_{(n-1),i} & \sum\limits_{i=1}^{n} X_{2,i} X_{n,i} \\[2ex] \cdot\cdot & \cdot\cdot & \cdot\cdot & \cdot\cdot & \cdot\cdot \\[2ex] \sum\limits_{i=1}^{n} X_{1,i} X_{(n-1),i} & \sum\limits_{i=1}^{n} X_{2,i} X_{(n-1),i} & \cdot\cdot & \sum\limits_{i=1}^{n} X_{(n-1),i}^{2} & \sum\limits_{i=1}^{n} X_{1,i} X_{(n-1),i} \\[2ex] \sum\limits_{i=1}^{n} X_{1,i} X_{n,i} & \sum\limits_{i=1}^{n} X_{2,i} X_{n,i} & \cdot\cdot & \sum\limits_{i=1}^{n} X_{1,i} X_{(n-1),i} & \sum\limits_{i=1}^{n} X_{n,i}^{2} \end{bmatrix}$$

$$(5.12)$$

In the matrix (5.12) the diagonal terms are the self-correlations and the others are the mutual correlations between the features.

Note

It is noted that when we have $n > q$ in Table 5.1, it is possible for a vector V to be outside the subspace of the feature vectors. In this space every vector is a superposition of the feature vectors and we can see this in the formula (5.3).

When we project the vector **V** onto the vectors of the feature we obtain the covariant components of the vector **V** as we can see in the formula (5.4). With the tensor $g_{i,j}$ we can obtain the contro-variant components from the covariant components by the expression (5.5). In conclusion we can compute a vector **V*** by the contro-variant components. Because **V** is outside the space generated by the feature space, we have that **V** is different from **V***.

It is easy to prove that among all possible vectors included in the space of the feature, V^* is the vector which differs from V by the minimum value. That is:

$$\| V - V^* \| = \text{minimum} \tag{5.13}$$

We outline the proof in this way:

Given a vector V that is not inside the three dimensional space (ψ_1, ψ_2, ψ_3). Every vector inside the subspace of the objects given by the feature space can be written as follows:

$$\mathbf{F} = w^1 \psi_1 + w^2 \psi_2 + w^3 \psi_3 \tag{5.14}$$

Given other vectors η_4, .., $\eta_{4\,n-3}$ in the space of the objects the generic vector V can be written in this way:

$$V = w^1 \psi_1 + w^2 \psi_2 + w^3 \psi_3 + \alpha^1 \eta_4 + \quad + \alpha^{n-3} \eta_{4\,n-3} \tag{5.15}$$

When we measure the distance, in this space of the objects, (that is an Euclidean space) we have:

$$D = \| V - F \| = \alpha_1^2 + \alpha_2^2 + \quad + \alpha_p^2 \tag{5.16}$$

Given a generic model F^* obtained without the tensor $g_{i,j}$ in the subspace of the feature , we have:

$$F^* = w_1{}^* \psi_1 + w_2{}^* \psi_2 + w_3{}^* \psi_3 \tag{5.17}$$

The distance between V and F^* is:

$$D^* = \| V - F^* \| = (w_1 - w_1{}^*)^2 + (w_2 - w_2{}^*)^2 + (w_3 - w_3{}^*)^2 \; \alpha_1^2 + \alpha_2^2 + \quad + \alpha_p^2 \tag{5.18}$$

We then have that $D \leq D^*$ where D is the minimum distance and F is the best model for theVector V.

Remark:

When n > q, we cannot always represent every vector as a superposition of the feature vectors. In this case the space of the features cannot be used as a reference space.

Given a set of n values V for the n objects, we cannot consider them as a superposition with different weights of the features. Even if we cannot decompose the V into its features, we can build a vector V* as superposition of the features the distance of which from the external vector V is the minimum.

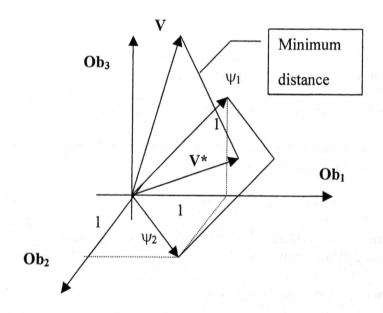

Figure 5.7. shows the distance between the external vector V and the vector V* inside the space of the features. When V* is calculated by the tensor g , the distance between V and V* is the minimum of all possible distances between V and one of the vectors inside the subspace of the features.

5.2.2 A Complete Example of the Three Dimensional Space of the Features

Given three features, n objects and one external vector V, we can build the Table 5.3.

Table 5.3. Table with three features, n objects and one external vector V.

	$\psi_1(t)$	$\psi_2(t)$	$\psi_3(t)$	V
Ob_1	X_1	Y_1	1	V_1
Ob_2	X_2	Y_2	1	V_2
..	.		..	......
Ob_{n-1}	X_{n-1}	Y_{n-1}	1	V_{n-1}
Ob_n	X_n	Y_n	1	V_n

For the previous table we have the tensor g:

$$
g_{i,j} = \begin{bmatrix} \sum_{i=1}^{n} X_i^2 & \sum_{i=1}^{n} X_i Y_i & \sum_{i=1}^{n} X_i \\ \sum_{i=1}^{n} X_i Y_i & \sum_{i=1}^{n} Y_i^2 & \sum_{i=1}^{n} Y_i \\ \sum_{i=1}^{n} X_i & \sum_{i=1}^{n} Y_i & n \end{bmatrix}
$$

When we change the variables X_i and Y_i into $x_i = X_i - <X_i>$ and $y_i = Y_i - <Y_i>$, (because $\Sigma\, x_i = 0$ and $\Sigma\, y_i = 0$) the tensor g can be written as follows:

$$
g_{i,j} = \begin{bmatrix} \sum_{i=1}^{n} x_i^2 & \sum_{i=1}^{n} x_i y_i & 0 \\[2ex] \sum_{i=1}^{n} x_i y_i & \sum_{i=1}^{n} y_i^2 & 0 \\[2ex] 0 & 0 & n \end{bmatrix}
$$

We note that the elements in the $g_{i,j}$ are very important statistical parameters. Here g_{11} is $nV(x)$ where $V(x)$ is the variance of x , g_{22} is $nV(y)$ where $V(y)$ is the variance of y and g_{12} is proportional to the correlation between x and y. For more statistical details see [16].

Because the vectors x_i and y_i are referred to the origin of the coordinates $\psi_1(t)$ and $\psi_2(t)$ it follows that g_{11} is the square of the module of the vector $\psi_1(t)$ and the same happens for g_{22}. When the two identities $\psi_1 = [x_1 , x_2 ,....,x_n]$ and $\psi_2 = [y_1 , y_2 ,....,y_n]$ are orthogonal, the feature space is a Euclidean space and $g_{12} = 0$. When the coordinates associated with the vectors ψ_1 and ψ_2 are dependent on each other then g_{12} is different from zero.

Considering the morphogenetic neuron, the different statistical parameters are collected in one geometric representation given by the space of the features.

To compute the covariant components we invert the previous matrices and obtain:

$$g^{i,j} = g_{i,j}^{-1} = \frac{1}{-\sum_{i=1}^{n} x_i^2 \sum_{i=1}^{n} y_i^2 + (\sum_{i=1}^{n} x_i y_i)^2} \begin{bmatrix} -\sum_{i=1}^{n} y_i^2 & \sum_{i=1}^{n} x_i y_i & 0 \\ \sum_{i=1}^{n} x_i y_i & -\sum_{i=1}^{n} x_i^2 & 0 \\ 0 & 0 & -\sum_{i=1}^{n} x_i^2 \sum_{i=1}^{n} y_i^2 + (\sum_{i=1}^{n} x_i y_i)^2 \end{bmatrix}$$

The covariant components are:

$$w_j = \frac{1}{-\sum_{i=1}^{n} x_i^2 \sum_{i=1}^{n} y_i^2 + (\sum_{i=1}^{n} x_i y_i)^2} \begin{bmatrix} -\sum_{i=1}^{n} y_i^2 \sum_{i=1}^{n} x_i V_i + \sum_{i=1}^{n} x_i y_i \sum_{i=1}^{n} y_i V_i \\ \sum_{i=1}^{n} x_i y_i \sum_{i=1}^{n} x_i V_i - \sum_{i=1}^{n} x_i^2 \sum_{i=1}^{n} y_i V_i \\ [-\sum_{i=1}^{n} x_i^2 \sum_{i=1}^{n} y_i^2 + (\sum_{i=1}^{n} x_i y_i)^2] \sum_{i=1}^{n} V_i \end{bmatrix}$$

The covariant components can also be written as follows:

$$w_j = \sum_{i=1}^{n} V_i \begin{bmatrix} \frac{1}{Det}[-\frac{\sum_{i=1}^{n} x_i V_i}{\sum_{i=1}^{n} f_i} \sum_{i=1}^{n} y_i^2 + \frac{\sum_{i=1}^{n} y_i V_i}{\sum_{i=1}^{n} f_i} \sum_{i=1}^{n} x_i y_i] \\ \frac{1}{Det}[\frac{\sum_{i=1}^{n} x_i V_i}{\sum_{i=1}^{n} V_i} \sum_{i=1}^{n} x_i y_i - \frac{\sum_{i=1}^{n} y_i V_i}{\sum_{i=1}^{n} V_i} \sum_{i=1}^{n} x_i^2] \\ 1 \end{bmatrix} = \sum_{i=1}^{n} V_i \begin{bmatrix} -\alpha_1 \sum_{i=1}^{n} y_i^2 + \alpha_2 \sum_{i=1}^{n} x_i y_i] \\ \alpha_1 \sum_{i=1}^{n} x_i y_i - \alpha_2 \sum_{i=1}^{n} x_i^2] \\ 1 \end{bmatrix}$$

The model is given by the expression:

$$F(x, y) = [(-\alpha_1 \sum_{i=1}^{n} y_i^2 + \alpha_2 \sum_{i=1}^{n} x_i y_i) x + (\alpha_1 \sum_{i=1}^{n} x_i y_i - \alpha_2 \sum_{i=1}^{n} x_i^2) y + 1] \sum_{i=1}^{n} f_i$$

where,

$$\alpha_1 = \frac{1}{\det} \frac{\sum_{i=1}^{n} x_i V_i}{\sum_{i=1}^{n} V_i}, \quad \alpha_2 = \frac{1}{\det} \frac{\sum_{i=1}^{n} y_i V_i}{\sum_{i=1}^{n} V_i}$$

$$C_x = \frac{\sum_{i=1}^{n} x_i V_i}{\sum_{i=1}^{n} V_i}, \quad C_y = \frac{\sum_{i=1}^{n} y_i V_i}{\sum_{i=1}^{n} V_i}$$

C_x and C_y are the centre of masses. We know that the function V with values V_1 , V_2 , .,V_n is a vector in the space of the objects. The function V is a new feature that we want to divide into the basic vectors $\psi_1(t)$, $\psi_2(t)$, $\psi_3(t)$. For simplicity we translate the values of the function V so as to have only positive values. When $\alpha_1 \neq 0$ and $\alpha_2 = 0$ the model becomes:

$$F(x, y) = [\alpha_1 ((-\sum_{i=1}^{n} y_i^2) x + \sum_{i=1}^{n} x_i y_i) y + 1] \sum_{i=1}^{n} f_i \qquad (5.19)$$

The slope of the previous model is independent of the centre of

mass C_x and is $m_{x,y} = -\dfrac{\sum_{i=1}^{n} y_i^2}{\sum_{i=1}^{n} x_i y_i}$ that is the slope of the regression

line x,y see [16].

When $\alpha_1 = 0$ and $\alpha_2 \neq 0$ the model becomes:

$$F(x,y) = [\alpha_2 (\sum_{i=1}^{n} x_i y_i) x - (\sum_{i=1}^{n} x_i^2) y) + 1] \sum_{i=1}^{n} f_i \qquad (5.20)$$

The slope of the previous model is independent of the centre of

mass C_y and is $m_{y,x} = \dfrac{\sum_{i=1}^{n} x_i y_i}{\sum_{i=1}^{n} x_i^2}$ which has the same slope as the re-

gression line y,x [16] . From J.P. Guilford we know that the regression equation may be used to predict the most likely measurement of one variable by the known measurement of another. Using the means of columns of a scattered diagram as the most probable corresponding Y values we predict Y's from the midpoint of interval X. Expressing this in another way, we are predicting the same Y value for a certain range of values of X. If we desire to be more accurate, we must be able to make predictions for all values of X. The regression line and the regression equation make this possible. The means of the columns tended to lie on a straight line, with some minor deviations from strict linearity. The line of the best fit for the straight line is the regression equation.

The linear model F(x,y) obtained with the feature space is the best model. The condition is a natural consequence of the building of the feature space. We do not use any particular processes to choose the best possible model. The best model is intrinsically connected with the space of the features. The function F(x,y) is the linear superposition of the two regression equations.

$$F(x,y) = (\sum_{i=1}^{n} f_i)[\alpha_1 r_1 + \alpha_2 r_2]$$

5.2.3 Change of Reference

Given a context which is identified by reference to a general coordinate, we will prove that there is always the possibility to change this reference or context and move to another context where the coordinates are Euclidean. In Figure 5.8 we represent in a schematic way the transition of different spaces into one Euclidean space.

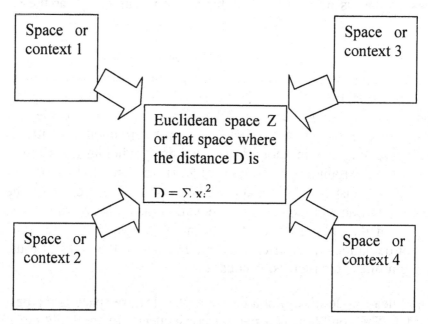

Figure 5.8. We show different contexts with different tensors g and the transformation of these contexts into one Euclidean flat space.

Proposition:

Every expression $\sum\limits_{i,j}^{n} g_{i,j}\psi_i\psi_j$ can be written in the quadratic form:

$$\sum_{i,j}^{n} g_{i,j} \psi_i \psi_j = (A_{11}\psi_1 + A_{12}\psi_2 + \ +A_{1n}\ \psi_n)^2 + (A_{22}\ \psi_2 +\ +$$

$$A_{2n}\psi_n)^2 + ..+(A_{nn}\ \psi_n)^2 = Z_1^2 + Z_2^2 +\ \ \ \ \ + Z_n^2 \tag{5.21}$$

where,

$$Z_1 = A_{11}\psi_1 + A_{12}\psi_2 +\ \ +A_{1n}\ \psi_n \tag{5.22}$$

$$Z_2 = A_{22}\ \psi_2 +\ \ +A_{2n}\psi_n$$

$$Z_n = A_{nn}$$

The coefficients $A_{i,j}$ can be obtained by using the Cholesky method.

Proof:

Given the quadratic form:

$$D(x_1,x_2,x_3) = g_{11}\ x_1^2 + g_{22}\ x_2^2 + g_{33}\ x_3^2 + 2\ g_{12}\ x_1\ x_2 + 2\ g_{13}\ x_1\ x_3 + g_{23}\ x_2 x_3$$

We can write this expression as:

$$D(x_1,x_2,x_3) = g_{11}\ x_1^2 + 2\ g_{12}\ x_1\ x_2 + 2\ g_{13}\ x_1\ x_3 + g_{22}\ x_2^2 + g_{33}\ x_3^2 + g_{23}\ x_2 x_3$$

And also as:

$$D(x_1,x_2,x_3) = g_{11}\ [x_1^2 + 2\ x_1\ [g_{12}\ x_2 + g_{13}\ x_3\ ] / g11] + g_{22}\ x_2^2 + g_{33}\ x_3^2 + g_{23}\ x_2 x_3$$

Because,

$$[x_1 + [g_{12} x_2 + g_{13} x_3] / (g11)]^2 = [x_1^2 + 2 x_1 [g_{12} x_2 + g_{13} x_3] / g11 + [g_{12} x_2 + g_{13} x_3]^2 / (g11)^2$$

The quadratic form can also be written as:

$$D(x_1,x_2,x_3) = g_{11} [x_1 + [g_{12} x_2 + g_{13} x_3] / g11]^2 - [g_{12} x_2 + g_{13} x_3]^2 / (g11) + g_{22} x_2^2 + g_{33} x_3^2 + g_{23} x_2 x_3$$

It is possible to write $D(x_1,x_2,x_3)$ as follows:

$$D(x_1,x_2,x_3) = [x_1\sqrt{g_{11}} + x_2 \frac{g_{12}}{\sqrt{g_{11}}} + x_3 \frac{g_{13}}{\sqrt{g_{11}}}]^2$$

$$+ \frac{g_{11}g_{22} - g_{12}g_{21}}{g_{11}} x_2^2 + \frac{g_{11}g_{33} - g_{13}g_{31}}{g_{11}} x_3^2 + 2 \frac{g_{11}g_{13} - g_{13}g_{21}}{g_{11}} x_2 x_3$$

For,

$$g_{11}^{\ 1} = \frac{g_{11}g_{22} - g_{12}g_{21}}{g_{11}} \ , \ g_{22}^{\ 1} = \frac{g_{11}g_{33} - g_{13}g_{31}}{g_{11}} \ , \ g_{12}^{\ 1} = \frac{g_{11}g_{13} - g_{13}g_{21}}{g_{11}}$$

we obtain,

$$D(x_1,x_2,x_3) = [x_1\sqrt{g_{11}} + x_2 \frac{g_{12}}{\sqrt{g_{11}}} + x_3 \frac{g_{13}}{\sqrt{g_{11}}}]^2 +$$

$$g_{11}^{\ 1} x_2^2 + g_{22}^{\ 1} x_3^2 + 2 g_{12}^{\ 1} x_2 x_3$$

But we can write:

$$D(x_1,x_2,x_3) = [x_1\sqrt{g_{11}} + x_2 \frac{g_{12}}{\sqrt{g_{11}}} + x_3 \frac{g_{13}}{\sqrt{g_{11}}}]^2 +$$

$$[x_2\sqrt{g_{11}^{\ 1}} + \frac{g_{22}^{\ 1}}{\sqrt{g_{11}^{\ 1}}} x_3]^2 + [x_3\sqrt{\frac{g_{11}^{\ 1}g_{22}^{\ 1} - g_{12}^{\ 1}g_{21}^{\ 1}}{g_{11}^{\ 1}}}]^2$$

When we substitute:

$$\sqrt{g_{11}^{\ 2}} = \sqrt{\frac{g_{11}^{\ 1}g_{22}^{\ 1} - g_{12}^{\ 1}g_{21}^{\ 1}}{g_{11}^{\ 1}}}$$

We obtain the transformation:

$$\begin{bmatrix} \sqrt{g_{11}} & \dfrac{g_{12}}{\sqrt{g_{11}}} & \dfrac{g_{13}}{\sqrt{g_{11}}} \\[2ex] 0 & \sqrt{g_{11}^{\ 1}} & \dfrac{g_{12}^{\ 1}}{\sqrt{g_{11}^{\ 1}}} \\[2ex] 0 & 0 & \sqrt{g_{11}^{\ 2}} \end{bmatrix} \begin{bmatrix} x_1 \\ x_2 \\ x_3 \end{bmatrix} = \begin{bmatrix} x'_1 \\ x'_2 \\ x'_3 \end{bmatrix}$$

The quadratic form of $D(x_1, x_2, x_3)$ in a non Euclidean space is transformed into the quadratic form:

$$D(x_1', x_2', x_3') = (x_1')^2 + (x_2')^2 + (x_3')^2$$

For a generic quadratic form in a non Euclidean space, the transformation by which we change the reference from the non Euclidean space into the Euclidean space is:

$$D(x_1, x_2, ..., x_n) = \sum_{i=1, j=1}^{n} g_{i,j} x^i x^j = [x_1 \sqrt{g_{11}} + x_2 \frac{g_{12}}{\sqrt{g_{11}}} + .. + x_n \frac{g_{1n}}{\sqrt{g_{11}}}]^2 + D(x_2, x_3, ..., x_n)$$

where $D(x_2, x_3, ..., x_n) = \sum_{j=2}^{n} \dfrac{g_{11}g_{jj} - g_{j1}g_{1j}}{g_{11}} x_j^2 + 2 \sum_{h=2, k=2}^{n} \dfrac{g_{11}g_{kh} - g_{k1}g_{1h}}{g_{11}} x_k x_h$

$$(5.23)$$

The general form of the transformation is:

$$
\begin{bmatrix}
\sqrt{g_{11}} & \dfrac{g_{12}}{\sqrt{g_{11}}} & \cdots & \dfrac{g_{1n-1}}{\sqrt{g_{11}}} & \dfrac{g_{1n}}{\sqrt{g_{11}}} \\[2ex]
0 & \sqrt{g_{11}^{\,1}} & \cdots & \dfrac{g_{1n-2}^{\,1}}{\sqrt{g_{11}^{\,1}}} & \dfrac{g_{1n-1}^{\,1}}{\sqrt{g_{11}^{\,1}}} \\[2ex]
\cdots & \cdots & \cdots & \cdots & \cdots \\[1ex]
0 & 0 & \cdots & \sqrt{g_{11}^{\,n-2}} & \dfrac{g_{12}^{\,n-2}}{\sqrt{g_{11}^{\,n-2}}} \\[2ex]
0 & 0 & \cdots & 0 & \sqrt{g_{11}^{\,n-1}}
\end{bmatrix}
\begin{bmatrix} X_1 \\ X_2 \\ \cdots \\ X_{n-1} \\ X_n \end{bmatrix}
=
\begin{bmatrix} X'_1 \\ X'_2 \\ \cdots \\ X'_{n-1} \\ X'_n \end{bmatrix}
\tag{5.24}
$$

The previous expression can be written as:

$\mathbf{U}\,\mathbf{x} = \mathbf{x'}$ and $\mathbf{g} = \mathbf{U}^T\,\mathbf{U}$

Using Cholesky's method [17] the transformation **U** can be obtained as follows:

$$
u_{i,i} = \sqrt{g_{i,i} - \sum_{k=1}^{i-1} u_{k,i}^2}\,, \qquad i = 1,2,...,n
$$

$$
u_{i,j} = \frac{1}{u_{i,i}}\left(g_{i,j} - \sum_{k=1}^{i-1} u_{k,i} u_{k,j}\right) \quad (j > i)
\tag{5.25}
$$

Example of Cholesky's method:

In (5.26) we show the matrix g and its transformation U by using Cholesky's method:

$$g = \begin{bmatrix} 2 & 1 & 1 \\ 1 & 3 & 2 \\ 1 & 2 & 2 \end{bmatrix} \text{ we have } U = \begin{bmatrix} \sqrt{2} & \dfrac{\sqrt{2}}{2} & \dfrac{\sqrt{2}}{2} \\ 0 & \dfrac{\sqrt{10}}{2} & \dfrac{3\sqrt{10}}{10} \\ 0 & 0 & \dfrac{\sqrt{15}}{5} \end{bmatrix} \tag{5.26}$$

the matrix **g** or the tensor $g_{i,j}$ is a symmetric tensor and can be factorized as follows:

$g = U^T U$. Here $A = U^T$ is the matrix that changes the reference from non Euclidean Space into a Euclidean Space x. The new variables x are independent. The vectors in x are linearly independent. Using the transformation A we can reduce the space to only independent variables.

5.2.4 Classification by Points in a General Space

Given n points P_k or objects in a general space of q dimensions with the feature q < n, we associate a value to any point of a function which is a vector in the space of objects or points. The Table 5.4 is used to generate the general space.

Table 5.4. Containing n points P_k in the q dimensional space of the features where every point has a value f_k.

	ψ_1	ψ_2	.	ψ_q	f
P_1	X_{11}	X_{21}	.	X_{q1}	f_1
P_2	X_{12}	X_{22}	....	X_{q2}	f_2
..	.		..	......	....
P_{n-1}	X_{1n-1}	$X_{2\,n-1}$	...	$X_{q(n-1)}$	f_{n-1}
P_n	X_{1n}	$X_{2\,n}$		X_{qn}	f_n

Proposition:

When we are in a Euclidean Space, we may assume every point with a mass with a value equal to the function f .The line which joins the centre of a mass with the origin of the axes is parallel to the vector that is perpendicular to the hyper-plane $\Sigma\ w^j\ \psi_j$.

Proof:

When g is an identity matrix, we have that for:

$$f = \Sigma\ w^j\ \psi_j\ ,\ w^j = w_j = \Sigma\ f_k\ \psi_{k,j}\ . \tag{5.27}$$

when we associate a mass equal to the value of the function f with every point, the centre of mass B is located in this point:

$$B_j = (\ \Sigma\ f_k\ \psi_{k,j}\ )\ /\ \Sigma\ f_k$$

The line that joins the centre of mass with the origin of the axes is given by the linear form:

$$\Sigma\ \eta^j\ \psi_j\ = 0\ ,\ \text{where}\ \eta^j\ = (\Sigma\ f_k\ \psi_{k,j}\ )\ /\ (\ \Sigma\ f_k\ \psi_{k,1})$$

but for Equation (5.26) and for the vectors V perpendicular to the linear form F we have:

$$\Sigma\ a_k\psi_k\ = 0$$

$$V = (1\ ,\ a_2\ /\ a_1\ ,\quad ..,a_n\ /\ a_1)$$

we note that the line that joins the centre of mass with the origin is perpendicular to the linear form $\Sigma\ w^j\ \psi_j\ = 0$.

For example, in Figure 5.9 we may show four points or objects in two dimensional space of the features. The same is shown in Table 5.5.

Table 5.5. In this table there are 4 points P_k, 2 features and one function.

	ψ_1	ψ_2	f
P_1	-1	1	1
P_2	1	1	2
P_3	1	-1	2
P_4	-1	-1	2

In accordance with the previous definition of centre of mass, the coordinates of the centre of mass are:

$B_1 = (-1 + 2 + 2 - 2) / (1 + 2 + 2 + 2) = 1 / 7$

$B_2 = (1 + 2 - 2 - 2) / (1 + 2 + 2 + 2) = - 1 / 7$

The line that joins the centre of mass with the origin, figure 5.9, is perpendicular to the following line in the feature space ψ_1, ψ_2.

$w^1 \psi_1 + w^2 \psi_2 = 0$

that we show again in Fingure 5.9.

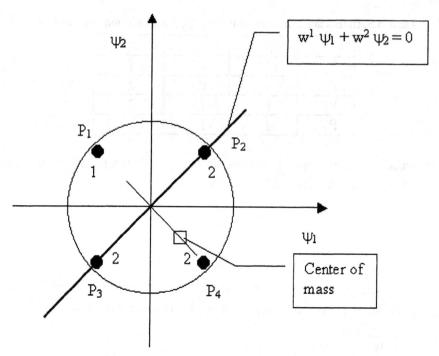

Figure 5.9. We show that in a Euclidean space the line that joins the centre of mass with the origin is perpendicular to the line $w^1 \psi_1 + w^2 \psi_2 = 0$.
The number in every point is the value of the function f in that point.

In Figure 5.9 we have three support points. One point with value 1
and two points with value 2 at the border of the set of points with
value 2. To solve the classification problem, we must find the line
L which is at the maximum distance between the border of the set
of points with value one and the border of the set of points with
value two contemporarily.

When we move the hyper –plane $\Sigma\, w^j\, \psi_j = 0$ in a parallel way to
obtain L as in the learning machine, we have the best separator between points with different masses and we solve the classification
problem shown in Figure 5.10.

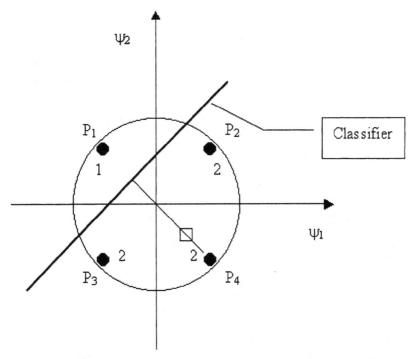

Figure 5.10. We show the line which solves the classification problem.

When we transform the coordinates of the features from a Euclid-ean space (ψ_1 , ψ_2) to a non- Euclidean space (ψ_1^* , ψ_2^*) , the points move to other points and the previous classifier is trans-formed into a classifier for the new points as shown in Figure 5.11.

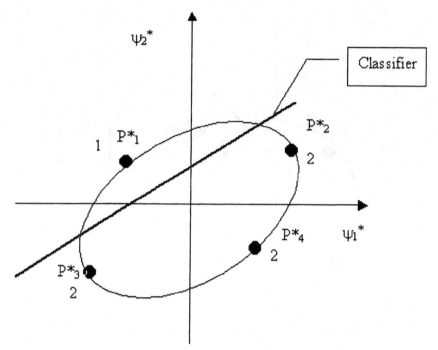

Figure 5.11. We show the image in Figure 5.10 obtained after a transformation from the Euclidean space (ψ_1, ψ_2) into the non Euclidean space (ψ_1^*, ψ_2^*).

5.2.5 Symmetry and Invariance in the Feature Space

As previously shown, every quadratic form $\sum\limits_{i=1, j=1}^{n} g_{i,j} x_i x_j$ can be written in the canonical form $D = \sum\limits_{i=1}^{n} X_i^2$. Every transformation for which D is invariant is a symmetric transformation. For other applications it is useful to know the form of the symmetric transformation. Every symmetric transformation $S\,X = Y$ has the property:

$$\sum_{i=1}^{n} X_i^2 = X^T X = (SX)^T (SX) = X^T S^T S X \text{ and } S^T S = \text{Identity matrix}$$

The product of the matrix S and its transpose S^T must equal the identity matrix. To find all possible matrices of the type S, we begin with the simple forms.

Proposition:

When the elements of S are contained in the set H = { -1 , 0 , 1 } every matrix S has one and only one element different from zero for each column and each row. In fact because $S^T S$ = the identity matrix, we must have $\sum_{i=1}^{n} s_{k,i} s_{i,h} = \delta_{k,h}$ where $\delta_{k,h}$ = 1 when k = h and $\delta_{k,h}$ = 0 when k ≠ h.

When we have only one element in S different from zero for every column and every row, the previous condition is satisfied.

For example

When n = 2 and because for every row and column we may have one and only one element different from zero in S, we have $2^2 + 2^2$ = 8 possible matrices. The possible matrices S in this case are:

$$\begin{bmatrix} 1 & 0 \\ 0 & 1 \end{bmatrix}, \begin{bmatrix} -1 & 0 \\ 0 & 1 \end{bmatrix}, \begin{bmatrix} 1 & 0 \\ 0 & -1 \end{bmatrix}, \begin{bmatrix} -1 & 0 \\ 0 & -1 \end{bmatrix}$$
$$\begin{bmatrix} 0 & 1 \\ 1 & 0 \end{bmatrix}, \begin{bmatrix} 0 & -1 \\ 1 & 0 \end{bmatrix}, \begin{bmatrix} 0 & 1 \\ -1 & 0 \end{bmatrix}, \begin{bmatrix} 0 & -1 \\ -1 & 0 \end{bmatrix}$$

(5.28)

Proposition

The number of symmetric matrices having the dimension n are $N = n\, 2^n$.

We note that the eight symmetric matrices are not independent one of the other. Only four matrices (generators) are independent one of the other and can generate all the other eight matrices. In our particular case the generators are:

$$\begin{bmatrix} 1 & 0 \\ 0 & 1 \end{bmatrix}, \begin{bmatrix} -1 & 0 \\ 0 & 1 \end{bmatrix}, \begin{bmatrix} 1 & 0 \\ 0 & -1 \end{bmatrix}, \begin{bmatrix} 0 & 1 \\ 1 & 0 \end{bmatrix}$$

Which generate the other matrices as follows:

$$\begin{bmatrix} -1 & 0 \\ 0 & 1 \end{bmatrix}\begin{bmatrix} 1 & 0 \\ 0 & -1 \end{bmatrix} = \begin{bmatrix} -1 & 0 \\ 0 & -1 \end{bmatrix}$$

$$\begin{bmatrix} -1 & 0 \\ 0 & 1 \end{bmatrix}\begin{bmatrix} 0 & 1 \\ 1 & 0 \end{bmatrix} = \begin{bmatrix} 0 & -1 \\ 1 & 0 \end{bmatrix}$$

$$\begin{bmatrix} 1 & 0 \\ 0 & -1 \end{bmatrix}\begin{bmatrix} 0 & 1 \\ 1 & 0 \end{bmatrix} = \begin{bmatrix} 0 & 1 \\ -1 & 0 \end{bmatrix}$$

$$\begin{bmatrix} -1 & 0 \\ 0 & 1 \end{bmatrix}\begin{bmatrix} 1 & 0 \\ 0 & -1 \end{bmatrix}\begin{bmatrix} 0 & 1 \\ 1 & 0 \end{bmatrix} = \begin{bmatrix} 0 & -1 \\ -1 & 0 \end{bmatrix}$$

Proposition

The product of two symmetric matrices is a symmetric matrix.

$$(S_1 S_2)^T (S_1 S_2) = S_2{}^T S_1{}^T S_1 S_2 = S_2{}^T \text{ Id } S_2 = \text{Identity}$$

Proposition

The matrix G(t) may be defined as follows:

$$G(t) = e^{St} = \sum_{n=0}^{\infty} \frac{t^n}{n!} S^n$$

Which is a symmetric matrix. It is easy to prove that $G(t)\,G(t)^T =$ Identity.

For example:

For $S = \begin{bmatrix} 0 & 1 \\ -1 & 0 \end{bmatrix}$, $G(t) = \begin{bmatrix} \cos(t) & \sin(t) \\ -\sin(t) & \cos(t) \end{bmatrix}$ $\qquad$ (5.29)

Proposition

The matrix,

$$G(t_1, t_2, \ldots, t_p) = \exp[S_1 t_1 + S_2 t_2 + \ldots + S_n t_n] = \exp[S_1 t_1]\,\exp[S_2 t_2] \ldots \exp[S_n t_n]$$

is a symmetric matrix.

For example:

When $n = 3$ we have the symmetric matrices:

$$S_1 = \begin{bmatrix} 0 & 1 & 0 \\ -1 & 0 & 0 \\ 0 & 0 & 0 \end{bmatrix}, \; S_2 = \begin{bmatrix} 0 & 0 & 1 \\ 0 & 0 & 0 \\ -1 & 0 & 0 \end{bmatrix}, \; S_3 = \begin{bmatrix} 0 & 0 & 0 \\ 0 & 0 & 1 \\ 0 & -1 & 0 \end{bmatrix}$$

where,
$$H = \exp(S_1\alpha + S_2\beta + S_3\gamma) = \exp(S_1\alpha)\,\exp(S_2\beta)\,\exp(S_3\gamma)$$

Which is a symmetric matrix $H\,H^T$ = identity

It is noted that because every metric $F = \sum_{i,j}^{n} g_{i,j}\psi_i\psi_j$ can be written in the quadratic form (see Section 5.2.2) we can give the symmetric transformation for every metric.

For example:

Given the metric

$F = 1.52\ X^2 + 0.52\ XY + 1.52\ Y^2 = (1.233\ X + 0.422\ Y)^2 + (1.158\ Y)^2 = Z_1^2 + Z_2^2$

Where,

$Z_1 = 1.233\ X + 0.422\ Y$ and $Z_2 = 1.158\ Y$ $\hspace{3cm}$ (5,30)

Because

$$\begin{bmatrix} \cos(\alpha) & \sin(\alpha) \\ -\sin(\alpha) & \cos(\alpha) \end{bmatrix}\begin{bmatrix} Z_1 \\ Z_2 \end{bmatrix} = \begin{bmatrix} Z_1\cos(\alpha) + Z_2\sin(\alpha) \\ -Z_1\sin(\alpha) + Z_2\cos(\alpha) \end{bmatrix}$$

Where, as we know from (5,29) for $t = \alpha$ the matrix is symmetric,

$[Z_1\cos(\alpha) + Z_2\sin(\alpha)]^2 + [-Z_1\sin(\alpha) + Z_2\cos(\alpha)]^2 =$
$= (Z_1^2 + Z_2^2)[\cos(\alpha)^2 + \sin(\alpha)^2] = Z_1^2 + Z_2^2$

But for (5,30),

$(1.233\ X^* + 0.422\ Y^*) = (1.233\ X + 0.422\ Y)\cos(\alpha) + (1.158\ Y)\sin(\alpha)$ $\hspace{2cm}$ (5.31)

$(1.158\ Y^*) = -(1.233\ X + 0.422\ Y)\sin(\alpha) + (1.158\ Y)\cos(\alpha)$

When we solve the system (5.31) we find the new coordinates X*
and Y* where the:

form $F = 1.52 \, X^2 + 0.52 \, XY + 1.52 \, Y^2$ is invariant.

Remark:

In the previous chapters the Adaptive Agent at different orders was
introduced. When we change the reference by means of a symmetric matrix S , the agent moves from one context to another (transformation at the second order). For the property of the symmetry,
the two contexts or references are different each from the other ,
but the metric $F = \Sigma \, g_{i,j} \, \psi^i \, \psi^j$ is the same. It is an invariant.

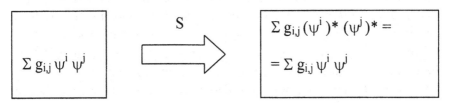

Illustration of a symmetric transformation

Figure 5.12. Adaptive second order agent which uses symmetric transformation S.

Here the distance expression and value (metric) are always the
same or invariant for the different contexts.

5.3 Deformation of the Space for Massive Data

In the previous approach we defined a set of features and objects
which are only a sample of the features. The structure was represented by Table 5.1. The data are mathematically expressed as a
sample set of vectors from the space of the objects. Each of these

vectors is a feature. The set of feature vectors generates the space of the features. This space is usually non Euclidean and is specified by a tensor g. Every vector V inside the feature space is a superposition of features. The coefficients of the superposition are the contro-variant components of the vector V in the feature space. With the tensor g and the vector V in the space of the objects we can compute the coefficients of the superposition.

In this context the data are divided into two parts. In one part there are the main data that specify the space where the context is built. The other part of the data are vectors in this space. It is necessary to research how to detect the main data which specify the environment or context. In general the search of the main data or basis vectors is an iterative process. We begin with a small set of basis vectors and we enlarge the set when we find vectors in the context which are not inside either the space of the basis vectors or that of the features. When the basis vectors are very large we accept that there are vectors not in these spaces. When the vector V is not in the space of these features, we can obtain the vector nearest the vector V inside the space of the features. In this way the basis functions are the fundamental entities to generate the best linear model of the vector V. In the previous chapters every object or sample has the same weight or mass. In this chapter we associate different weights to the objects. The set of weights is denoted $P = \{ p_1, p_2$, $p_n \}$.

The new data obtained by the multiplication of the values of an object (point) by the weight that belongs to P change the reference (space of features) or context of the Adaptive Agent. In Figure 5.13 we show the change of the context and reference by the weights.

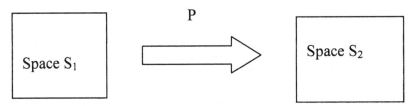

Figure 5.13. Second order adaptive agent which uses the set of weights P to move from one context to another.

In Table 5.6 we show both Table 5.1 and the weights P.

Table 5.6. Set of data and weights.

	ψ_1	ψ_2	.	ψ_q
Ob_1	$p_1 X_{11}$	$p_1 X_{21}$	.	$p_1 X_{q\,1}$
Ob_2	$p_2 X_{12}$	$p_2 X_{22}$	....	$p_2 X_{q\,2}$
..	.		..	......
Ob_{n-1}	$p_{n-1} X_{1n-1}$	$p_{n-1} X_{2\,n-1}$	...	$p_{n-1} X_{q\,(n-1)}$
Ob_n	$p_n X_{1n}$	$p_n X_{2\,n}$		$p_n X_{q\,n}$

In Figure 5.14 we show an example of weighted points in the space of the feature in two dimensions:

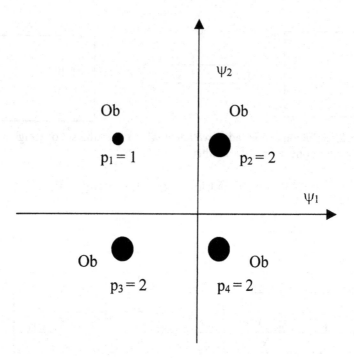

Figure 5.14. Weighted points.

With the new data in Table 5.6 we can compute the tensor g as follows:

$$
g_{i,j} = \begin{bmatrix}
\sum_{i=1}^{n}(p_i Z_{1,i})^2 & \sum_{i=1}^{n} p_i^2 Z_{1,i} Z_{2,i} & .. & \sum_{i=1}^{n} p_i^2 Z_{1,i} Z_{(n-1),i} & \sum_{i=1}^{n} p_i^2 Z_{1,i} Z_{n,i} \\
\sum_{i=1}^{n} p_i^2 Z_{1,i} Z_{2,i} & \sum_{i=1}^{n}(p_i Z_{2,i})^2 & .. & \sum_{i=1}^{n} p_i^2 Z_{2,i} Z_{(n-1),i} & \sum_{i=1}^{n} p_i^2 Z_{2,i} Z_{n,i} \\
.. & .. & .. & .. & .. \\
\sum_{i=1}^{n} p_i^2 Z_{1,i} Z_{(n-1),i} & \sum_{i=1}^{n} p_i^2 Z_{2,i} Z_{(n-1),i} & .. & \sum_{i=1}^{n}(p_i Z_{(n-1),i})^2 & \sum_{i=1}^{n} p_i^2 Z_{1,i} Z_{(n-1),i} \\
\sum_{i=1}^{n} p_i^2 Z_{1,i} Z_{n,i} & \sum_{i=1}^{n} p_i^2 Z_{2,i} Z_{n,i} & .. & \sum_{i=1}^{n} p_i^2 Z_{1,i} Z_{(n-1),i} & \sum_{i=1}^{n}(p_i Z_{n,i})^2
\end{bmatrix}
$$

$$(5.32)$$

Example:

Given the Boolean functions:

$Y_1 = X_1$ AND NOT (X_1), $Y_2 = X_1$ AND X_2, $Y_3 = X_1$ OR X_2 (5.33)

And the set of data in Table 5.7:

Table 5.7. Space of the Boolean functions.

Objects	ψ_1	ψ_2
Ob_1	-1	-1
Ob_2	1	-1
Ob_3	1	1
Ob_4	-1	1

With the data in the Table 5.7 we can compute the tensor g, that is:

$$g = \begin{bmatrix} 4 & 0 \\ 0 & 4 \end{bmatrix}$$

When in the Boolean function the result is zero, the weight is one, but when the result of the Boolean function is one, the weight is two. So for the Boolean functions Y_1, Y_2 and Y_3, we have the set of weights P_1, P_2 and P_3 in the Table 5.8:

Table 5.8. Space of the Boolean functions and the weight sets P_1, P_2, P_3, for the functions in (5.33).

Objects	$\psi_1 = X_1$	$\psi_2 = X_2$	P_1	P_2	P_3
Ob_1	-1	-1	1	1	1
Ob_2	1	-1	1	1	2
Ob_3	1	1	1	2	2
Ob_4	-1	1	1	1	2

Because the function $Y_1 = X_1$ AND NOT (X_1) is always false, the weights in P_1 are always equal to the minimum value which is one. By means of (5.32) and the weights in P_1 it is possible to compute the tensor g_1 the value of which is:

$$g_1 = \begin{bmatrix} 4 & 0 \\ 0 & 4 \end{bmatrix}$$

Because the function $Y_1 = X_1$ AND X_2 is true only when X_1 and X_2 are true, the weights in P_2 are equal to the maximum value two when $\psi_1 = 1$, $\psi_2 = 1$ (object Ob_3). In the other cases the weights are equal to one. By means of (5.32) and the weights in P_2 it is possible to compute the tensor g_2 the value of which is:

$$g_2 = \begin{bmatrix} 7 & 3 \\ 3 & 7 \end{bmatrix}$$

The function Y_1 always has the values equal to zero. All the other functions have at least one value equal to one. So the tensor g_1 that represents the function Y_1 has the minimum values. We denote the function Y_1 as the reference function. To compare the tensor of the functions with the reference, we subtract the generic tensor g from the tensor g_1:

In Y_2, we subtract the tensor g_1 from g_2 and we obtain:

$$g_2 - g_1 = \begin{bmatrix} 3 & 3 \\ 3 & 3 \end{bmatrix}$$

With a change of scale with factor 3, we obtain:

$$\frac{g_2 - g_1}{3} = \begin{bmatrix} 1 & 1 \\ 1 & 1 \end{bmatrix} \qquad\qquad (5.34)$$

Function Y_3

$$g_3 - g_0 = \begin{bmatrix} 9 & -3 \\ -3 & 9 \end{bmatrix}$$

With a change of scale with factor 3, we obtain:

$$\frac{g_2 - g_1}{3} = \begin{bmatrix} 3 & -1 \\ -1 & 3 \end{bmatrix} \tag{5.35}$$

We note that in (5.35) the values in the principal diagonal are equal to the number of values equal to one in Y_3.

The function Y_3 is equal to one for the values:

$Y_3 (X_1 , X_2) = 1$

for $(X_1, X_2) = (1, 0)$, $(X_1, X_2) = (0, 1)$, $(X_1, X_2) = (1, 1)$

For the Table 5.8 the previous values correspond to $(1, -1)$ of the object Ob_2, $(-1,1)$ of the object Ob_4 and $(1,1)$ of the object Ob_3. The values of (5.35) are given by the sum of the following products:

	1, -1			-1, 1			1, 1		
1	1 -1	+	-1	1 -1	+	1	1 1	=	3 -1
-1	-1 1		1	-1 1		1	1 1		-1 3

The weighted averages of the values of the features ψ_1 and ψ_2 are as follows.

For the function Y_1 and the weights in P_1 we have:

$< \psi_1 > = [(1) (-1) + (1) (1) + (1) (1) + (1) (-1)] / 4 = 0$

$< \psi_2 > = [\, (\, 1 \,) \, (\, -1 \,) + (\, 1 \,) \, (\, -1 \,) + (\, 1 \,) \, (\, 1 \,) + (\, 1 \,) \, (\, 1 \,) \,] / 4 = 0$

For the function Y_2 and the weights in P_2 we have:

$< \psi_1 > = [\, (\, 1 \,) \, (\, -1 \,) + (\, 1 \,) \, (\, 1 \,) + (\, 2 \,) \, (\, 1 \,) + (\, 1 \,) \, (\, -1 \,) \,] / 4 = 1/4$

$< \psi_2 > = [\, (\, 1 \,) \, (\, -1 \,) + (\, 1 \,) \, (\, -1 \,) + (\, 2 \,) \, (\, 1 \,) + (\, 1 \,) \, (\, 1 \,) \,] / 4 = 1/4$

For the function Y_3 and the weights in P_3 we have:

$< \psi_1 > = [\, (\, 1 \,) \, (\, -1 \,) + (\, 2 \,) \, (\, 1 \,) + (\, 2 \,) \, (\, 1 \,) + (\, 2 \,) \, (\, -1 \,) \,] / 4 = 1/4 =$
$(\, 1 + 1 - 1 \,) / 4$

$< \psi_2 > = [\, (\, 1 \,) \, (\, -1 \,) + (\, 2 \,) \, (\, -1 \,) + (\, 2 \,) \, (\, 1 \,) + (\, 2 \,) \, (\, 1 \,) \,] / 4 = -1/4$
$= (-1 + 1 - 1 \,) / 4$

The weighted average value multiplied by $2^2 = 4$ gives the sum of the values of the attributes, without weights, for which the previous functions assume the value equal to one. In conclusion the average value gives useful information on the value of the functions Y_1 Y_2 Y_3. The average value gives the position of the origin of the feature space.

Important remark

Between the Boolean function and the weighted feature space in table 5.8 there exists a connection for which we can associate one weighted feature space (context) with every Boolean function. In this way the second order agent generates its context by its Boolean functions.

For the Boolean functions with three variables we have the data in Table 5.9.

Table 5.9. Values of the variables X_1, X_2 and X_3 for a Boolean function in three dimensions.

	$\psi_1 = X_1$	$\psi_2 = X_2$	$\psi_3 = X_3$
Ob_1	-1	-1	-1
Ob_2	1	-1	-1
Ob_3	1	1	-1
Ob_4	-1	1	-1
Ob_5	-1	-1	1
Ob_6	1	-1	1
Ob_7	1	1	1
Ob_8	-1	1	1

For the functions:

$$Z_1 = \text{NOT } X_1 \text{ AND } X_1 \text{ and } Z_2 = X_1 \text{ AND } (X_2 \text{ XOR } X_3) \qquad (5.36)$$

we have the weights P_1 and P_2 in table 5.10.

Table 5.10. Values of the variables X_1, X_2 and X_3 for a Boolean function and weights for the functions Z_1 and Z_2.

	$\psi_1 = X_1$	$\psi_2 = X_2$	$\psi_3 = X_3$	P_1	P_2
Ob_1	-1	-1	-1	1	1
Ob_2	1	-1	-1	1	1
Ob_3	1	1	-1	1	2
Ob_4	-1	1	-1	1	1
Ob_5	-1	-1	1	1	1
Ob_6	1	-1	1	1	2
Ob_7	1	1	1	1	1
Ob_8	-1	1	1	1	1

For the formulas 5.36 and 5.32 for the function Z_1, we have the following:

$$g_1 = \begin{matrix} 8 & 0 & 0 \\ 0 & 8 & 0 \\ 0 & 0 & 8 \end{matrix}$$

and for the function Z_2 we have the following:

$$g_2 = \begin{matrix} 14 & 0 & 0 \\ 0 & 14 & -6 \\ 0 & -6 & 14 \end{matrix}$$

We note that the function Z_1 has the minimum number of values in g_1. All the other functions have at least one value equal to one. So the tensor g_1 that represents the function Z_1 is the basic reference. To compare the tensor of the functions with the reference, we subtract the generic tensor g from the tensor g_1 In Z_2 we subtract the tensor g_1 from g_2 and we obtain:

$$g_2 - g_1 = \begin{matrix} 6 & 0 & 0 \\ 0 & 6 & -6 \\ 0 & -6 & 6 \end{matrix}$$

when we change the scale with factor 3 we obtain:

$$(g_2 - g_1) / 3 = \begin{matrix} 2 & 0 & 0 \\ 0 & 2 & -2 \\ 0 & -2 & 2 \end{matrix} \qquad (5.37)$$

The function Z_2 is equal to one for the values:

$$Z_2 (X_1 , X_2 , X_3) = 1$$

for $(X_1 , X_2 , X_3) = (1 , 1 , -1)$, $(X_1 , X_2 , X_3) = (1 , -1 , 1)$

The values of (5.37) are given by the sum of the following products:

Product	1	1	-1		Product	1	-1	1
1,	1	1,	-1	+	1,	1	-1,	1
1,	1	1	-1		-1,	-1	1	-1
-1	-1	-1	1		1	1	-1	1

$$(5.38)$$

The 5.37 can be obtained by the values (5.38) generated by the function Z_2. We can repeat the same considerations also for Boolean functions with more variables.

5.4 Conclusion

Research on neural and neuro-fuzzy architectures indicates there is a need for a more general concept than that of the *neural unity*, or *node*, introduced in the pioneering work of McCulloch and Pitts. The neural unity widely used today in artificial neural networks can be considered as a non-linear filter. From this basic unity so-called "integrated" neural architectures may be built, where many different neural networks cooperate. In order to do research on such neural architectures a descriptive language is needed in which an artificial neural network can be considered as a single, dedicated entity. On the basis of these considerations we propose a generalization of the concept of neural unity.

The model developed in this chapter is denoted as a Morphogenetic Neuron .We use the name "neuron" because the activation function of such a device is characterized, in the same way as in classical neural unities, by a *bias potential* and by a *weighted sum* of suitable, in general non-linear, functions or basis functions. The attribute "morphogenetic" was chosen because the data determine the space which generates the models by the linear operation of superposition, see (5.3). Using the morphogenetic neuron it becomes possible to automatically produce a network of conventional neural unities implementing a given input-output transfer function, *without* the need to resort to laborious methods such as supervised training. This concept of the morphogenetic neuron appears to coincide with recent research in neurobiology. For example the auditory place cells are believed to 'compute' the azimuth position (left - right) and elevation position (up - down) of a sound source. The position is obtained by measuring the Inter-Neural Time difference (ITD) between signals which are delayed with respect to each other due to the difference in path-lengths between the sound source and each ear. In this case two inputs control the same auditory place cells. The two signals are *fused or superposed* to obtain values of the azimuth and elevation positions. Classically, data fusion or superposition is motivated by the presumption that high quality primary data are of paramount importance to the success of *all* subsequent signal processing stages. The concept of the morphogenetic neuron shows that this does not always need to be the case.

We introduced a novel generalization of the concept of neural unity, which has been named *morphogenetic neuron*. From an abstract point of view, it is a generic analog input-output device through which two elementary operations are possible: the "Write" and "Read" operation, *i.e.*

(1) The operation (*"Write"*), starting from suitable *reference basis functions* generates the space of the features.
(2) The computation operation (*"Read"*), which is started from a given function as a vector in the feature space can be used in the space previously generated to produce a linear model of the given functions.

The advantages of the introduction of morphogenetic neurons can be listed as follows:

(a) Once a suitable way to implement the Write and Read operations has been found, we have an analog device which is able to satisfy given constraints on input data *without the need to introduce either logical operators in an explicit way, or to use training procedures based on known examples*;

(b) The concept of the morphogenetic neuron allows us to unify, within a single theory, concepts belonging to very different research domains. That is all the different types of neural unities may be considered as particular kinds of morphogenetic neurons. For example even holography can be seen as being equivalent to the operation of a particular morphogenetic neuron. In this case an *interference* process which gives rise to the formation of the hologram corresponds to the WRITE operation. The *diffraction* process which produces the object's image corresponds to the READ operation;

(c) Since the morphogenetic neuron is essentially analog, its operation depends on suitable continuous quantities. We may apply standard methods of dynamical systems theory [18] or use the framework of the field theory as formulated in theoretical physics [19]. The classical representation of the neural network models, on the contrary, preclude a direct application of these methods.

(d) By introducing a suitable approximation for the READ operation, it is possible to obtain automatically the design for a suitable network architecture consisting of conventional neural unities. These would implement the input-output transfer function of a given morphogenetic neuron having to resort to training for example. It is clear that these advantages depend for their success on a careful definition of the WRITE and READ operations.

Bibliography

1. McCullough W. and Pitts W.H., "A logical calculus of the ideas immanent in nervous activity", Bull. Math. Biophys, vol.5, pp.115-133, 1943.

2. Nobuo S., "The extent to which Biosonar information is represented in the Bat Auditory Cortex", in: "Dynamic Aspects of Neocortical Function", ed. G.M. Edelman, John Wiley, New York, 1984.

3. Resconi G., "The morphogenetic Neuron" in Computational Intelligence : Soft Computing and Fuzzy.- Neuro Integration with Application, Springer NATO ASI Series F Computer and System Science, vol.162, pp.304-331, 1998. Editors Okyay Kaynak, Lotfi Zadeh, Burhan Turksen, Imre J.Rudas.

4. Resconi G. and Pessa E., Poluzzi R., "The Morphogenetic Neuron", *Proceedings fourteenth European meeting on cybernetics and systems research*, pp.628-633, April 14-17, 1998.

5. Murre J.M.J., *Learning and categorization in modular neural networks*, Erlbaum, Hillsdale, NJ, 1992.

6. Benjafield J.G., *Cognition*. Prentice-Hall, Englewood Cliffs, NJ, 1992.

7. Salinas E. and Abbott L.F., "A model of multiplicative neural responses in parietal cortex", Proc. Natl. Acad. Sci. USA, vol.93, pp.11956-11961, 1996.

8. Duif A.M. and van der Wal A.J., "Enhanced pattern recognition performance of the Hopfield neural network by orthogonalization of the learning patterns", Proc. 5th Internat. Parallel Processing symposium, Newport (CA), 1991.

9. Salinas E. and Abbott L.F., "Invariant Visual responses From Attentional gain Fields", The American Physiological Society, pp.3267-3272, 1997.

10. Resconi G., van der Wal A.J. and Ruan D., "Speed-up of the MC method by using a physical model of the Dempster-Shafer theory", Int. J. of Intelligent Systems, Special issue on FLINS'96, vol. 13, no. 2/3, pp 221-242, 1998.

11. Resconi G. and van der Wall A.J., Morphogenetic neural networks encode abstract rules by data, Information Sciences, vol.142, pp. 249-273, 2002.

12. Alexandre Pouger and Lawrence H. Snyder, Computational approaches to sensorimotor transformations, Nature Neuroscience –supplement, vol.3, pp.1192-1198, November 2000.

13. Maximilian Riesenhuber and Tomaso Poggio, Models of object recognition, transformations, Nature Neuroscience –supplement, vol.3, pp.1199-1204, November 2000.

14. Alexandre Pouger and Lawrence H. Snyder, Computational approaches to sensory-motor transformations ,are contained in the Nature Neuroscience supplement to vol.3, November 2000.

15. Bernhard Scholkopf, *Learning with Kernels and Alexander J.Smola*, The MIT Press Cambridge, Massachusetts, London England, 2002.

16. Guilford J.P., *Fundamental Statistics,* McGraw-Hill, International Student Education, New York, 1965.

17. Sidney Yakowitz and Ferenc Szidarovszky, *An introduction to numerical computations*, second edition, Macmillan Publishing Company, New York, 1989.

18. John Guckenheimer and Philip Holmes, *Nonlinear Oscillations, Dynamical Systems, and Bifurcations of Vector Fields*, Springer Verlag, New York, 1983.

19. Muirhead H., *The Physics of Elmentary Particles*. Pergamon Press, Oxford 1965.

Chapter Six

Logic Actions of Adaptive Agents

6.1 Introduction to Modal Logic

There is a general acceptance that Kripke models (Chellas, 1980) [1] present the fundamental semantics for modal logic. There are several generalisations of Kripke models. Among these generalisations there are the so-called Scott-Montague models (Chellas, 1980). These have the most general characteristics of such models. In this section we present a brief description of Kripke and Scott-Montague models.

Let ATOMS be elements of a denumerable set of *atomic sentences.* Let L_{ML} be a language for modal logic formed from ATOMS using propositional connectives such as ¬ (negation), ∧ (conjunction), ∨ (disjunction), → (material implication), and ↔ (equivalence) and modal operators such as □ (necessity) and ◊ (possibility).

A *Kripke model* is defined as the following three-tuple:

$M = <W, R, V>,$

Where W is a non-empty set of possible worlds,

$R \subseteq W \times W$ is an accessibility relation on W,

V is a valuation operator for atomic sentences at each possible world, $w \in W$:

$V: \text{ATOM} \times W \to \{T, F\}$

Where ATOM are individual sentences and W is the set of worlds. The evaluation function V by means of the individual sentence and the particular world gives the logic value True (T) or False (F).

The valuation operator is extended in the usual way and in particular, for modal operators, $\Box$ necessity and $\Diamond$ possibility. We then have:

$$V(\Box p, w) = T \Leftrightarrow \forall w'(wRw' \Rightarrow V(p, w')) = T \Leftrightarrow \Xi_w \subseteq \|p\|^M,$$

$$V(\Diamond p, w) = T \Leftrightarrow \exists w' (wRw' \text{ and } V(p, w')) = T \Leftrightarrow \Xi_w \cap \|p\|^M \neq \varnothing$$

For the previous expressions one sentence is necessarily true when it is true in all accessible worlds and one sentence is possibly true when it is true in almost one accessible world. With the symbol $\|p\|^M$ we denote the set of worlds where in the model M the sentence p is true. The symbol $\Xi_w = \{w' \mid wRw'\}$ and $\|p\|^M = \{w \mid V(p, w) = T\}$. Note that $\|p\|^M$ is called the truth set associated with the sentence of an atomic sentence p in a model M in the literature of philosophical logic. In what follows, we omit the superscript 'M' unless confusion arises. It is well-known that different conditions on accessibility relations correspond to different axiom schemas.

In conclusion the modal logic with the definition of the possible world, explains the literal meaning of possibility and necessity. So we enlarge the classical logic with new operators and with a more flexible logic evaluation of the sentences connected with a possible world. This can be associated to the agent belief or point of view.

6.2 Meta-theory based upon Modal Logic

In a series of papers Klir, *et al.*, (1994 [6], 1995 [7]); Resconi, *et al.*, 1992, 1993, 1996) [11,12,13,14] a meta-theory was developed. The new approach is based on Kripke model of modal logic. A Kripke model is given by the structure:

$M = < W, R, V >$.

Resconi, *et al.* (1992-1996) suggested that a further function Ψ should be added:

$\Psi : W \rightarrow R$

Where R is the set of real numbers assigned to the worlds in W in order to obtain the new model:

$$S1 = < W, R, V, \Psi > \tag{6.1}$$

That is for every world there is an associated real number. With the model S1, we can build the *hierarchical meta-theory*. Here we can calculate the expression for the membership function of "truth" using fuzzy set theory to verify a given sentence. This is done using a computational method based on {1,0} values. These correspond to the truth values {T , F} assigned to a given sentence as the response of a world, a person, or a sensor for example.

At this point we must ask: "What are the linkages between the concepts of a population of observers, a population of possible worlds and the algebra of fuzzy subsets in a well defined universe of discourse?" To restate this question, we need to point out that the fuzzy set theory was introduced to take account of the imprecision found in natural languages. However this imprecision means that a word representing an "entity" such as temperature and velocity for example cannot have a crisp logic evaluation. The meaning of a word in a proposition can usually be evaluated in different ways for different assessments of an entity by the various agents such as worlds.

An important principle is:

"We cannot separate the assessments of an entity without any loss of property in the representation of that entity itself".

Different and in some cases conflicting evaluations for the same proposition may be given for the same entity. The presence of conflicting properties within the entity itself is the principal reason for the imprecision in the representation of the entity. For example, suppose the entity is a particular temperature of a room. We then ask for the property *cold*, when we have no instrument to measure that property. Under these circumstances the meaning of the entity, "temperature", is composed of assessments that are the *opinions* of a population of observers who evaluate the predicate "cold". Without the population of observers, and their assessments, we cannot have the entity "temperature" and the predicate "cold". When we move from crisp set to the fuzzy set, we move from atomic *elements*, which are the individual assessments to non-atomic *entities*, or aggregate assessments. In an abstract way, the population of these assessments of the entity becomes the population of the worlds.

We associate the set of propositions with crisp logic evaluation to every possible world. Perception based data embedded in a context generate imprecise properties. Using the perception based data, as assessments of an entity, we are able to evaluate the properties of an entity. The evaluation can contain conflicting elements. In such cases, the world population assessment composed of individual assessments gives the context structure of the perception based data. If we know only the name of a person (entity), we cannot know if s/he is "old". Additional observation-based information that s/he is married, is a manager, and that s/he plays with toys is conflicting information of the proposition that associates her/him with the predicate "old".

The aim of this book is to show that by using a model of agents (worlds) that with their perception generate the basic imprecision process , we can both simplify the definitions of the operations in fuzzy logic and expose and explain deeper issues embedded within the fuzzy theory.

Consider a sentence such as: "John is tall" where $x \in X$ is a person. "John" is in a population of people X and "tall" is a linguistic term of a linguistic variable, the height of people. Let the meta-linguistic expression that represents the proposition "John is tall" be written in fuzzy set canonical form as follows:

$p_A(x)::="$ $x \in X$ isr $A"$.

Here "isr" means " $x \in X$ is in relation to a fuzzy information granule A", and $p_A(x)$ is the proposition that relates x to A. Next consider the imprecise sentence "John is tall" can be modeled by different entities, such as different measures of the physical height h. For each height h, let us associate the opinions, based on their perceptions, of a set of observers or sensors that give an evaluation for the sentence "John is tall". Any observer or sensor can be modeled in an abstract way by a world $w_k \in W$, where W is the set of all possible worlds that compose the indivisible entity It should be noted that the world, which may be a person or the sensor, does not say anything about the qualification, that is even a descriptive gradation such as "John's being tall", but it just verifies on the basis of a particular valuation scheme.

Having dealt with these preliminaries, we can now write $p_A(x)$ where $V(p_A(x), w_k)$ evaluates T as true. We now assign $\eta(w_k, x, A) = 1$ if $V(p_A(x), w_k) = T$ or $\eta(w_k, x, A) = 0$ if

$V(p_A(x), w_k) = F$ or false. With this background we next define the membership expression of truth for a given atomic sentence in a finite set of worlds $w_k \in W$ as follows:

$$\mu_{p_A}(x) = \frac{(set \; of \; worlds \; \text{where} \; p_A(x) \; is \; true)}{|W(x)|} = \frac{\sum_k \eta(w_k, x, A)}{|W(x)|} \quad (6.2)$$

In the equation (6.2) any value of the variable x is associate with the set of the worlds $W(x)$ which represents the opinion of the population of the observers, based on their perceptions. Where p_A represents the proposition that the atomic sentence "John is tall", and A = "tall" such that $V(p_A(x)), w_k) = T$ for x = "John". Here $|W|$ is the cardinality of the set of worlds in our domain of concern. *Recall once again that these worlds, $w_k \in W$, may be agents, sensors, persons, etc.* Let us define the subset of worlds $W_A = \{ w_k \in W \,|V\, (p_A(x), w_k) = T\}$. We can write expression (6.2) as follows:

$$\mu_{p_A}(x) = \frac{|W_A|}{|W|} \quad (6.3)$$

With the understanding that W_A represents the subset of the worlds W where the valuation of $p_A(x)$ is "true" in the Kripke sense. For the special case, where the relation R in the Kripke model is:

$$w_k \, R \, w_k$$

any such world w_k is isolated from all others. The membership expression is computed as the value of Ψ in S1 stated in (6.1) above. It is computed using the expression:

$$\Psi = \frac{1}{|W|}$$

for any (single) world w in W. Thus we can write the expression (6.2) as follows:

$$\mu_{p_A}(x) = \sum_k \eta(w_k)\Psi_k \tag{6.4}$$

It should be noted that the valuation operator V assigns {T, F} to every atomic sentence p_A for any world. In the different expressions (6.2), (6.3), (6.4), we assume that every evaluation in one world is the result of a perception of the same object from different points of view. These different perceptions generate the impossibility to assign one logic value to the predicate of the object. The multi logic values of the object generate confusions and uncertainties.

6.3 Adaptive Agents Having Different Order and Semantic Events

We know from the previous chapters that an adaptive agent of the second order transforms one set of rules from one context into another. We also know that the set of rules which come from one context C_1 into another context C_2 is not always equal to the set of rules in the context C_2. This generates confusion and conflict. In Figure 6.1 we show a context which receives different projections from other contexts. Here we have different rules that generate confusion. In fact we know that a coherent context has one and only one set of rules.

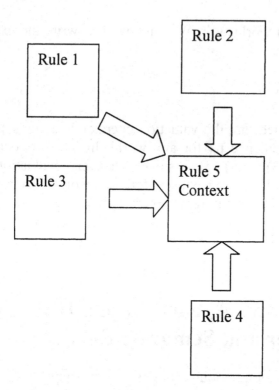

Figure 6.1. Illustrating that many different rules from different contexts may arrive at a context. As every context has only one set of rules, all the other rules generate uncertainty, ambiguity or fuzziness.

To represent the conflicting situation shown in Figure 6.1 we associate adaptive agents of different orders. We consider a space $S = (X_1, X_2, , X_N)$ which is divided into a set of finite units that define the sub-domains of S. Each unit identifies a possible world.

The space S is the space of possible projections of the rules into the context. Every world is a projection of one set of rules into the context. In Figure 6.2 we show the worlds.

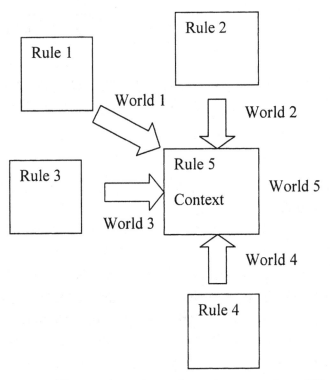

Figure 6.2. Different rules arrive at the same context from different contexts. Every channel that transports the rules is a possible world.

Let these worlds be labeled as W_i where the index i identifies the unit X_i in the space S. Such a scenario of possible worlds is represented by a context (framework) where agents (experts) operate as evaluators. In the language of the adaptive agents, the agents in the context evaluate if the various sets of rules, which come from different contexts, are the same or are different. We assume that the judgment or the evaluation of the agents is a simple judgment as "to whether the rules are true rules." This latter judgment is our logic evaluation of the proposition p.

At the beginning we have only a dichotomy judgment. Given a proposition p, each agent evaluates p to be "true" or "false" in a world, W_i.

In order to explain this framework, we next define a "Semantic Elementary Event", SEE, and assign a "Semantic Measure", SM, to each SEE.

Definition 1:

An evaluation of a proposition p, in a possible world W_i, is called a "Semantic Elementary Event".

Definition 2:

When the logic value of the proposition p in a possible world W_i within S is "true", then the "Semantic Measure", SM, is defined as: $SM((W_i,p) = \eta_i(p)$, where $\eta_i(p) \in R$. When the logic value is "true":

$$\sum_{i=1}^{N} \eta_i(p) \leq N.$$

In other words, when a proposition p is evaluated to be "true" in a possible world W_i, then $\eta_i(p)=1$ and thus $SM(W_i,p)=1$ and therefore when p is "false", $\eta_i(p) = 0$ and thus $SM(W_i,p)=0$.

It is to be noted that Definition 2 is for only one W_i. Since it is possible that there could be a number of W_i's in a universe W, we need to assess the truth given to a proposition for all W_i's. That is, we need to determine the truth assigned to a proposition for all W_i's in agreement with (6.2) and (6.3) given above.

In this regard, in the meta-theory and within the context of a universe W the Semantic Measure of a proposition, SM (W, p), is equal to the value of the membership function of the truth assigned to a given proposition p, that is $SM(W, p)=\mu_p(x)$.

In other words, when any "Semantic Elementary Event" has the same Semantic Measure in a set of worlds, then we have the "Semantic Measure" for a set of worlds where p is true. This is given by the expression:

$$SM(W, p) = \frac{\text{Number of Semantic Elementary Events}}{N} = \frac{|W_i|}{|W|}$$

For the meta-theory we have:

$$\mu_p(x) = \frac{\text{Number of Semantic elementary events}}{N} = \frac{|W_p|}{|W|}$$

Definition 3:

The Semantic Measure of a proposition p in the universal set W of the possible world $W_i \in W$ where i=1,...,N, is equal to "one" if all possible worlds were to respond "Yes" to a proposition p:

$$SM_i = SM(W_i, p) = \eta_i(p) \qquad \text{and}$$

$$SM(W, p) = (|W|)^{-1} \Sigma^N_{i=1} \eta_i(p) = (|W|)^{-1} \sum_{i=1}^{N} SM(W_i, p) = 1,$$

$$SM(W, p) = 1.$$

It can be seen in this situation all the sets of rules which in Figure 6.2 reach the context are equal and so we have:

p = "the rules which are the true rules"

is always true in all the worlds or channels that transport a set of rules from the external to the internal part of the context.

Thus, any set of possible worlds can be associated with a Semantic Measure that has either a positive or zero value.

Therefore, given the sets of possible worlds B_i, and B_j with the property that $B_i \cap B_j = \emptyset$, we have:

$$SM(B_i \cup B_j) = SM(B_i) + SM(B_j)$$

Thus SM is an additive function with a positive value and it is completely additive.

Summing up, in the meta-theory, the Semantic Measure, SM, is in general equal to the value of the fuzzy membership function $\mu_p(x)$ of truth in the fuzzy set theory (Resconi and Türkşen, 2001) [16], that is, $SM(p)=\mu_p(x)$. Furthermore, the Semantic Measure has the same properties as ordinary probability.

6.3.1 Example of Possible Worlds

Suppose we have a context which receives rules from 63 other contexts. The simple rule used in our context is the proposition p = "the square is black". It is true when the square is black and false when the square is white. Next, suppose the large square shown in Figure 6.3 is the total image of all the rules that arrive from the 64 contexts and where we include our proper context. Because in one context we may use only one rule, the presence of many different rules generates confusion and uncertainty.

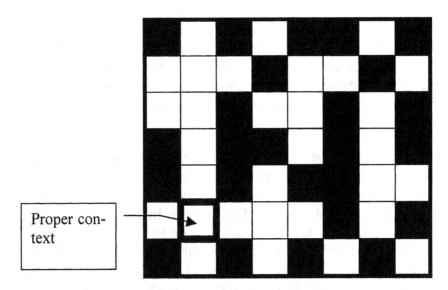

Figure 6.3. An observation of a local cell (Proper context), or particular rule , may cause us to infer a conclusion on the color of the local cell which when applied to the whole square is incoherent.

When we cannot experiment and/or observe the whole big square (object), and we have only local and/or partial information about the small squares, we may hypothesize that the big square is either black or white. That is if we observe only that a little square is black then we may think that all the squares are black, and if we observe that a little square is white, then we may think that all the squares are white. When observations of this kind are followed by inferred thinking, they are a source of uncertainty expressed by the modal logic concept of a world that in this case is associated with a little square.

Using the information based on the color of a little square we may infer the color of the big square. Thus when the little square is black then we hypothesize that the big square is black. When the little square is white then we hypothesize that the big square is white. For the hypothetical process of generalization from the little square to an unknown big square, we generate errors and uncertainties. These are the result of the conflicting situation arising from the conceptual granulation in little squares or worlds. Under these circumstances when we extend the local properties of one part of the object to the whole of the object, we make a non-rational extension, with the possibility of generating conflicting situations and uncertainty.

The definition of possible worlds comes from this inappropriate extension. In fact, the concept of dividing an object into granules or worlds from which agents can make a hypothetical assessment for the whole object is the source of difficulties. In the contexts of most empirical sciences, we usually are unable to use clearly justified extensions. For example, in Physics, Galileo hypothesized that the gravitational acceleration is constant. Galileo extended this hypothesis to the whole Universe near the earth. But later, Newton discovered that the acceleration is not constant but changes with the inverse of the square of the distance. Then Newton's "law of the square inverse of the distance" was again extended to the whole Universe. Last century Einstein found that in some parts of the Universe Newton's law is not correct. In conclusion because we cannot obtain all the data of all the bodies in the universe, our only possibility is to extend our hypothesis in an arbitrary way. We use our local knowledge of the world and extend it to the whole Universe with the resulting degree of uncertainty.

As the adaptive agent can use only one set of rules in a particular context, it assumes that every set of rules found is correct in every context. For the adaptive agent it is necessary to limit the information to one context to have a coherent situation where the rules do not change.

6.4 Models of Semantic Measure with the Fuzzy Logic Operations

In this section we develop semantic measure concepts and their formulas for "AND", and "OR" combinations of any two propositions within the context of Type 1 fuzzy set theory where the membership function is crisp (G. Klir an Bo Yuan) [38]. There is, however, another set of semantic measure models and their formulas which need to be developed within the context of Interval-Valued Type 2 theory where the membership function is fuzzy. We will leave this for the future.

6.4.1 The Operation AND

We investigate combinations of the truths for any two propositions, such as ' "John is tall" is true' AND '"John is heavy" is true'. Given that we know the truth membership functions of $\mu_{p_1}(x)$ for P_1: '"John is tall" is true', and that $\mu_{p_2}(x)$ for P_2: '"John is heavy" is also true'. It is clear from Section 3 that if we know the set of possible worlds $W_1 = \{W_i\}$ where p_1 is true and the set of possible worlds $W_2 = \{W_j\}$ where p_2 is true, then we can compute:

$$SM(p_1) = \mu_{p_1}(x) \quad \text{and} \quad SM(p_2) = \mu_{p_2}(x) \quad , \quad \text{that is,}$$
$$\mu_{p_1}(x) = |W_1| / |W| \quad \text{and} \quad \mu_{p_2}(x) = |W_2| / |W|.$$

The details of this calculation were discussed in Resconi and Türkşen (2001) [16]. The formulas developed in that paper will be reviewed to build the Semantic Measure.

Given that $|W_1| = N_1$ is the number of Semantic Elementary Events in which the proposition p_1 is true and $|W_2| = N_2$ is the number of Semantic Elementary Events in which the proposition p_2

is true in any world. We now generate a new event p such that "p₁ and p₂" is "true" is given by the expression:

$$p = p_1 \wedge p_2 .$$

Here "and" is interpreted as being equivalent to the "∧" operation.

For Semantic Elementary Events that have the same Semantic Measure value, we can write:

$$SM(p) = \mu_p(x) = \mu_{p_1 \wedge p_2}(x) = \frac{|W_1 \cap W_2|}{|W|}$$

Because the SM has the same formal properties as the probability measure, we can write its expression in terms of the conditional measure as:

$$SM(p_1 \wedge p_2) = SM(p_1/p_2)SM(p_2) = \mu_{p_1/p_2}\mu_{p_2} = \frac{|W_1 \cap W_2|}{|W|}$$

where $\mu(p_1/p_2)$ is a new function denoted conditional truth membership function. For the dissonant case when $W_1 \supseteq W_2$, that is, $\mu_{p_1}(x) \geq \mu_{p_2}(x)$, we show that the previous expression can be written in the same form as the fuzzy logic AND operation in the following way:

$$\mu_{p_1 \wedge p_2} = \mu_{p_1/p_2}\mu_{p_2} = \min(\mu_{p_1}, \mu_{p_2}) - \mu_{p_2}[1 - \mu_{p_1/p_2}]$$

This formulation is true and we have proved it previously [16]. In the meta-theory when $\mu_{p_1}(x) \geq \mu_{p_2}(x)$ we can write:

$$\mu_{P_1 \wedge P_2}(x) = \min[\mu_{P_1}(x), \mu_{P_2}(x)] - \frac{|cW_1 \cap W_2|}{|W|},$$

where $\mu_{P_1}(x) = \dfrac{W_1}{W}$, $\mu_{P_2}(x) = \dfrac{W_2}{W}$

and,

$$\mu_{P_2}[1 - \mu_{P_1/P_2}] = \frac{|cW_1 \cap W_2|}{|W|}$$

such that,

$$\mu_{P_1/P_2} = 1 - \frac{|cW_1 \cap W_2|}{|W|} \frac{1}{\mu_{P_2}}.$$

Therefore, we have:

$$\mu_{P_1/P_2} = 1 - \frac{\mu_{P_1 \wedge P_2} - \min(\mu_{P_1}, \mu_{P_2})}{\mu_{P_2}}$$

When $W_2 \subseteq W_1$ that is, the consonant case, we have that $\mu_{P_1/P_2} = 1$ and:

$$\mu_{P_1 \wedge P_2} = \mu_{P_2} = \min(\mu_{P_1}, \mu_{P_2}).$$

On the other hand, when,

$$\mu_{P_1/P_2} = \mu_{P_1} = 1 - \frac{|cW_1 \cap W_2|}{|W|} \frac{1}{\mu_{P_2}}$$

or,

$$\frac{\left| cW_1 \cap W_2 \right|}{|W|} = [1 - \mu_{p_1}]\mu_{p_2},$$

We get $\mu_{p_1 \wedge p_2} = \mu_{p_1}\mu_{p_2}$ and the two semantic measures or truth membership functions are independent.

Example:

Using the results obtained so far, we have the algebraic product as the "AND" combination for the case of the independent events as:

$$\mu_{p_1 \wedge p_2} = \mu_{p_1/p_2}\mu_{p_2} = \mu_{p_1}\mu_{p_2}$$

Whereas we have the bounded difference that is, Lukasiewicz Conjunction, for when the events are not independent as:

$$\mu_{p_1 \wedge p_2} = \mu_{p_1/p_2}\mu_{p_2} = \max[0, \mu_{p_1} + \mu_{p_2} - 1]$$

This can be written as:

$$\mu_{p_1 \wedge p_2} = \max[0, \mu_{p_1} + \mu_{p_2} - 1] - \min(\mu_{p1}, \mu_{p2}) + \min((\mu_{p1}, \mu_{p2}) =$$
$$= \min(\mu_{p1}, \mu_{p2}) - [\min(\mu_{p1}, \mu_{p2}) - \max[0, \mu_{p_1} + \mu_{p_2} - 1]$$

From this, we derive the definition of the conditional membership function for the bounded difference as:

$$\mu_{p_1/p_2} = 1 - \frac{[\min(\mu_{p1}, \mu_{p2}) - \max[0, \mu_{p_1} + \mu_{p_2} - 1]}{\mu_{p_2}}$$

when $\mu_{p_1} \geq \mu_{p_2}$.

In Figure 6.4, we show the conditional membership function for various values of $\mu_{p_1}, \mu_{p_2} \in [0,1]$ for the bounded difference, (Lukasiewicz Conjunction):

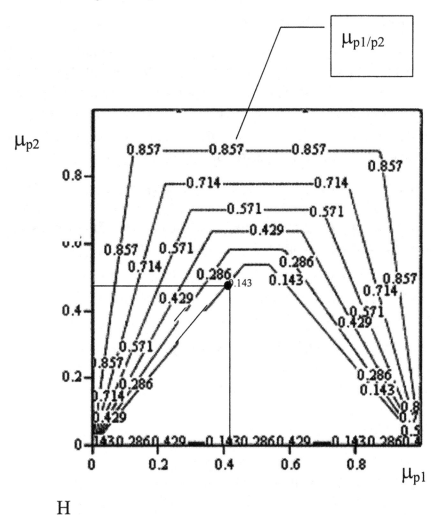

H

Figure 6.4. Values of the conditional membership μ_{p_1/p_2} for different values of μ_{p1} and μ_{p2} for the bounded difference.

It should be noted that at the border $\mu_{p_1/p_2} = 1$ for the bounded difference $\mu_{p_1 \wedge p_2} = \max(0, \mu_{p_1} + \mu_{p_2} - 1)$.

6.4.2 The Operation OR

For the OR combination, let us next investigate the combination of the truth for any two propositions, such as ' "John is tall" is true' OR ' "John is heavy" is true', given that we know the truth membership functions of $\mu_{p_1}(x)$ for p_1 ' "John is tall" is true', and $\mu_{p_2}(x)$ for p_2: ' "John is heavy" is true'. We have the semantic measure analogous to the usual probability measure as:

$$SM(p_1 \vee p_2) = SM(p_1) + SM(p_2) - SM(p_1 \wedge p_2)$$

When we rewrite the same equation in the context of the worlds we have the relation among the fuzzy truth measures as:

$$SP(p_1 \vee p_2) = \mu_{p_1 \vee p_2} = \frac{|W_1 \cup W_2|}{|W|}$$

which can be written in the following way:

$$\mu_{p_1/p_2} = \mu_{p_1} + \mu_{p_2} - \mu_{p_1 \wedge p_2} = \mu_{p_1} + \mu_{p_2} - \mu_{p_1/p_2} \mu_{p_2}$$

For $\mu_{p_1}(x) \geq \mu_{p_2}(x)$ we can write;

$$\mu_{p_1 \vee p_2} = \mu_{p_1} + \mu_{p_2} - [\min(\mu_{p_1}, \mu_{p_2})] + \frac{|cW_1 \cap W_2|}{|W|}$$

$$= \max(\mu_{p_1}, \mu_{p_2}) + \frac{|cW_1 \cap W_2|}{|W|}$$

or alternately,

$$\mu_{p_1 \vee p_2} = \mu_{p_1} + \mu_{p_2} - \mu_{p_2}[1 - \frac{|cW_1 \cap W_2|}{|W|}]$$

6.4.3 The Operation NOT and Operator Γ under the Action of an Agent

Given a set of worlds with a parallel evaluation of the true value for every world, we can use an operator Γ to change the logic value of the propositions in the worlds.

In the modal logic we have the possible $\Diamond$ and necessary $\Box$ unary operators that change the logic value of the proposition p. One proposition p in one world w is necessarily true or $\Box$ p is true, when p is true in all accessible worlds. One proposition p in one world w is possibly true or $\Diamond$ p is true when it is true at least in one world accessible from w. We add a new operator Γ near these operators.

Definition:

The operator Γ is unary and represents the action of one agent on the proposition p in one world:

$$\Gamma(p, w) = (\neg p, w) \quad \text{or} \quad \Gamma(p, w) = (p, w)$$

Given the special world w, the agent can change the proposition in this world from p to $\neg$ p or the agent cannot change the proposition p. The operator Γ represents the distribution of the changes of the proposition p inside the set of worlds.

With the agent operator Γ we can generate any type of t-norm or t-conorm that are the several possible AND and OR in the fuzzy set theory defined in the book of G. Klir and Bo Yuan [38] and also non – classical complement operators, see again G. Klir and Bo Yuan [38].

Given a set of worlds and the proposition p defined in all the worlds, for the (6.2) the complement operator is $1 - \mu$. But when we introduce the operator Γ into some worlds, the agent does not change p but in other worlds it changes p into $\neg$ p. So the set of worlds is divided into two parts. Γ breaks the traditional symmetry in the modal logic for which one proposition p is defined in the same way in all the worlds even if the logic value of p is different in the different worlds. In modal logic one proposition cannot be defined as p in one world and $\neg$ p in another. This is the main difference between classical logic and fuzzy logic. When the operator Γ breaks the classical symmetry the complement is not equal to the traditional value $1 - \mu$ but we can have new and more flexible complement operations in fuzzy set theory. All this will be clarified in the following examples.

Example:

Given five worlds (w_1 , w_2 , w_3 , w_4 , w_5) two propositions p and q and the logic evaluation in the five worlds:

V (p) = (True, True, False, False, False) and   V (q) = (False, True, True, True, False)   (6.5)

From (6.2), we have $\mu_p = \dfrac{2}{5}$ and $\mu_q = \dfrac{3}{5}$

The evaluation of V (p OR q) = (True, True, True, True, False) where we use the classical logic operation in every world. So we have:

$$\mu_{p \, or \, q} = \frac{4}{5} > \max \left(\frac{2}{5} , \frac{3}{5} \right) = \frac{3}{5} \qquad\qquad (6.6)$$

For the classical logic, p is the same in all the worlds or in a more simple way:

$(p \text{ in } w_1 , p \text{ in } w_2 , p \text{ in } w_3, p \text{ in } w_4 , p \text{ in } w_5)$

and the symmetry is respected. But when we introduce an external agent and its action Γ is to break the classical symmetry, we have

$(\Gamma p \text{ in } w_1 , \Gamma p \text{ in } w_2 , \Gamma p \text{ in } w_3, \Gamma p \text{ in } w_4 , \Gamma p \text{ in } w_5) =$
$(\neg p \text{ in } w_1, p \text{ in } w_2 , \neg p \text{ in } w_3, p \text{ in } w_4 , p \text{ in } w_5)$

and the evaluation under the action of Γ is the following

$$V (\Gamma.p) = (\text{False, True, True, False, False}) \qquad\qquad (6.7)$$

The new evaluation is still an evaluation for the same proposition p but from another point of view. So in the traditional logic given one world we have only one possible evaluation for the same proposition, but in the new fuzzy logic many different evaluations are possible under the action of different agents. This opens a new way to give a logic interpretation of uncertainty and of vague propositions in the natural language. Using the evaluation in (6.7) obtained by the operator Γ and the evaluation in (6.5) for the proposition q we obtain another evaluation for the same expression:

p OR q .

So we have

$$V (p \text{ OR } q) = (\text{False, True, True, True, False})$$

And,

$$\mu_{\Gamma p \text{ or } q} = \frac{3}{5} = \max (\frac{2}{5} , \frac{3}{5}) = \frac{3}{5}$$

The agent has transformed the general fuzzy logic operation OR (6.6) into the Zadeh max rule. The agent Γ changes the evaluation for the same logic expression and this gives a justification for the different t-norm and the t-conorm in the fuzzy set theory.

We know that given the proposition p and the membership function μ_p the Zadeh complement or operator NOT is $\mu_{cp} = 1 - \mu_p$. For Zadeh minimum rule we have $\mu_{p\ AND\ cp} = \min(\mu_p, 1 - \mu_p)$ that in general is different from zero. So the absurd expression p AND cp that in classical logic is always equal to zero or false can be different from zero in the fuzzy logic.

Given the evaluation:

V (p) = (True, True, False, False, False) and V (cp) = (False, False, True, True, True)   (6.8)

We have:

V (p AND cp) = (False, False, False, False, False)

We are in the classical condition for which the absurd expression is always false in all the worlds. But when we introduce the agent operator and we break the symmetry we may have:

$(\Gamma$ p in w_1 , Γ p in w_2 , Γ p in w_3, Γ p in w_4 , Γ p in w_5) = $(\neg$ p in w_1 , $\neg$ p in w_2, p in w_3, $\neg$ p in w_4 , $\neg$ p in w_5)

So we have:

V (Γcp) = (True, True, True, False, False) and
V (p AND Γcp) = (True , True , False , False , False),

$$\mu_{p \text{ and } \Gamma c\, p} = \min\left(\frac{2}{5}, \frac{3}{5}\right) = \frac{2}{5} \text{ which is a value different from}$$

zero or false.

We remark that $V(\Gamma cp) = V(c\,\Gamma p)$ and $\Gamma\Gamma p = p$.

Using the operator Γ the NOT operator can be changed in such a way so as to have either the Zadeh complement or the Sugeno complement. Other complements can also be obtained by the use of Γ operator.

We know that the use of the operator Γ is, for the adaptive agent, the operator that adjusts a set of rules in such a way as to obtain complete agreement or is also able to change the set of rules so as to obtain a situation where the incoherence increases.

In fact in Figure 6.2 we associate five worlds with the channels. In the situation of complete coherence we have this evaluation for the rules

$$V(p) = (\text{True, True, True, True, True}) \tag{6.9}$$

Where $p =$ "the rules which are true rules". In the fuzzy or incoherent situation by using the same proposition p we can have this evaluation

$$V(p) = (\text{True, True, True, False, False}) \tag{6.10}$$

In this case two channels transport to the context rules that are incoherent with the others (see Figure 6.2). With the action of one agent we can have this situation:

$$(\Gamma\, p \text{ in } w_1, \Gamma\, p \text{ in } w_2, \Gamma\, p \text{ in } w_3, \Gamma\, p \text{ in } w_4, \Gamma\, p \text{ in } w_5) =$$
$$(p \text{ in } w_1, p \text{ in } w_2, p \text{ in } w_3, \neg p \text{ in } w_4, \neg p \text{ in } w_5) \tag{6.11}$$

We move from the coherent situation in (6.9), to the incoherent situation (6.10). But within the same situation (6.11) we move from the incoherent situation (6.10) to the coherent situation (6.9).

In Figure 6.5 we show the action of the operator Γ:

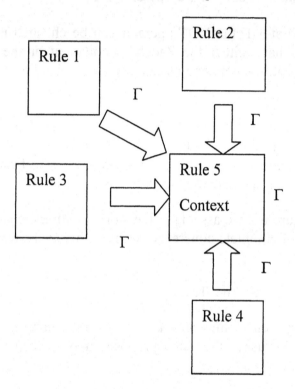

Figure 6.5. Illustrating how the operator Γ can change the degree of incoherence within the context.

6.5 Symmetry and Break of Symmetry

Consider a universe S that consists of the set of n worlds:

$S = \{w_1, w_2, ., w_n\}$

Suppose, in each world, or alternatively local context, we evaluate the proposition p and generalize the evaluation for all possible worlds in S. We then have a global symmetry in S where for *every local world or context, we always have the proposition p.* Every global symmetry is associated with an invariant proposition. The invariant proposition p is an evaluation that is always the same in any world, situation, position, point of view. For example in physics a conservative system always has the same energy for all the geometric translations of the system. The proposition:

p = "The system has the energy E"

is always the same in a conservative system for every transformation or change of the world.

In the meta-theory of uncertainty (G. Resconi, G. Klir, 1992) [11] every proposition p in one world may be considered to be valid in all the other worlds having the same global properties or symmetry. In this case, no conflicts exist between local worlds and the universe S when the proposition has the same logic values in all of the worlds. In fact, when p is true in one world, then it is true also in all the other worlds in S. There are no conflicts existing between the local judgment in one world and the global judgment for all worlds. In the classical logic, for a given universe S and for every world in it, the logic value is always either true or false. In many cases, we have only local information and local true or false information. But we then generalize this local information and the logic value of the proposition p to all the other worlds in S. We can consider the set S to be the superposition of all the worlds and the evaluation of the object S to have the average value of the evaluations in each of the worlds. Thus when we have symmetry, the average value is the logic value which is either true or false for all the possible worlds in S.

If the symmetry is broken, in one world we define the proposition p, in another world we define the proposition ¬ p. Thus in one case the logic value of S is true and in the other case the logic value of S is false. Hence S is in the superposition of two logic states such that one is p and the other is ¬ p. For the same object S , p can be true and false at the same time. This is comparable with the Schrodinger's cat in quantum mechanics, where it cannot be determined whether the cat is dead or alive. For the classical logic the cat cannot be alive and dead at the same point and time. But for the fuzzy logic it is possible. For the definition of the operator Γ when p = "the cat can be alive" and ¬ p. "the cat can be dead", the attribute of the cat is the superposition of p and ¬ p so it can be alive and dead at the same time and point.

Every observer in a local world destroys the coherence among the worlds in S when in the local world it appears that only one of the true or false states occurs for all the worlds. In the quantum computer, we denote a state to be true or false with a qubit. The term qubit means quantum beat and is associated with a quantum state that represents the True or False situation. In this case we have only a probability that one particle is in one state or in another. No sure rules are known to define position, velocity or other properties of the particles. So we have a set of conflicting rules of the particles and only probability rules for all the worlds (states).

This type of misjudgment of state can be found also in Galileo's explanation of the movement of the Earth around the Sun. Galileo saw with the telescope that Jupiter has many satellites, which move around it. Because in the mind of Galileo all parts of the universe move by the same rule, he generalized that if it is true that the satellites move around Jupiter, the earth must also move around the sun in a similar way. However Jupiter with its satellites is one world or a part of the universe and the earth with the sun is another world or a part of the universe. Thus if it is the case there is a dynamical symmetry in the universe where if a property is true in one part, it is true in every part of the universe. Galileo made the correct inferential reasoning. He could not be sure of the symmetry in the universe at the time when he made the generalization and because of this he could not and should not have generalized the property of Jupiter and its planets to all the other parts of the Universe. We know that Galileo was lucky and made the correct generalization from a part of the solar system to the whole of the solar system and to the universe.

In conclusion, when we generalize from local information to the object universe S, the object could have conflicting properties if there is no symmetry. In such cases the average value of the logic values of the proposition p, given in a part of the universe S, is a fractional value or membership value of the objects S.

It is assumed that when making a generalization from a world to all the others we can move without limitation.

6.5.1 Symmetry Break, SB Operator and Operator Γ

In a particular world given the proposition p and the operator Γ we can now define the Symmetry break operator SB(p) as:

SB(p) = False, when $\Gamma p = p$, and when SB(p) = True, we have $\Gamma p = \neg p$

When in a world there is no break of symmetry, that is, SB(p) = False, then in this world the proposition p does not change. When SB(p) = True, we then have a break of symmetry. The proposition p, in a particular world then changes from p into its negation $\neg$ p. In conclusion the generic world transformation operator Γ can be written as follows:

$$\Gamma p = (p \wedge \neg SB(p)) \vee (\neg p \wedge SB(p)). \tag{6.12}$$

However if we know that the proposition p changes, we can then calculate the value of the break of symmetry S, by the expression.

$$SB(p) = (\neg p \wedge \Gamma p) \vee (p \wedge \neg \Gamma p)$$

Example:

Given $S = \{w_1, w_2, w_3, w_4, w_5\}$ and $V(p) = \{$True, False, True, False, True$\}$, the operator Γ changes the logic value of $V(p)$:

$V(\Gamma p) = \{$True, True ,True, False, False $\}$

We know that $\Gamma \Gamma p = p$. In fact in the world where SB = True we have $\Gamma p = \neg p$. and $\Gamma [\Gamma p] = \neg \Gamma p = p$ When in the world SB = False we have $\Gamma p = p$. and $\Gamma [\Gamma p] = \Gamma p = p$. In any case $\Gamma \Gamma p = p$.

Example:

Consider the proposition $p = $ "George, who is 1.70 m high is tall ". Consider now a universe of five worlds which may be persons, agents or contexts.

$S = \{w_1, w_2, w_3, w_4, w_5\}$

Suppose we have that in all five worlds $V(p) = $ True, that is, all five agents assign a "True" value to the proposition p, then the valuation of p is given as:

$V(p) = \{$True, True, True, True, True$\}$.

In such a case there is a complete accord or symmetry among the persons in S. That is S has a symmetric set of worlds. Suppose for the world number 3, we have $SB(p) = 1$. In this case we have a break of symmetry. That is the 3rd agent assigns a "False" value to

p. In fact for the expression (6.12) in the world 3, we obtain Γ p = $\neg$ p. Here we have a new evaluation which is:

V(p) = {True, True, False, True, True}

Thus in this last case the symmetry between local and global evaluation is broken and we cannot generalize. Because in the third world V(p) = False we must state that for all S the proposition p is false. On the other hand, because there exists a world where p is also true, we could also state that for all S the proposition p is true. Because the five judgments are to be superposed for the same person George, this break of symmetry generates imprecision. Therefore, the membership value, for the value x = 1.70m in the fuzzy set "George is tall " has to be 4/5.

6.5.2 Sugeno Complement

Given the Sugeno complement:

$$c(a) = \frac{1-a}{1+\lambda a}, \qquad\qquad (6.13)$$

where a is the fuzzy logic variable (membership function), c(a) is the complement of the fuzzy logic variables and λ is a parameter that defines complement operator.

We remark that when $\lambda > 0$, we have c(a) $\leq$ 1 – a . When a = 0 c(a) = 1 and when a =1 c(a) =0. For the boundary condition SB = "false" in all the worlds that change in synchronism. The worlds are strictly dependent on one and the other and the set of all the worlds form a unique coherent entity.

In the other cases we calculate the difference of c(a) with 1 – a , which has:

$$S(a) = c(a) - (1-a) = \frac{\lambda a(a-1)}{1+\lambda a}$$

We divide the set of worlds in W_s into two types. In one type FT there are the worlds for which:

p = "False" and SB = "True"

and in the other type TT there are the worlds where:

p = "True" , SB = "True".

Proposition:

The difference $\Omega = \frac{2N_{TT}}{N} - \frac{N_{FT}}{N}$ has the property c(Ω) = Ω and

$$(1-a) + \Omega = (1-a) + \frac{2N_{TT}}{N} - \frac{N_{FT}}{N} = c(a)$$

Where in Ω, N_{TT} is the number of worlds of the type TT , N_{FT} the number of worlds of the type FT and N is the total number of worlds.

Proof:

For Sugeno complement, we have:

$$\Omega = \frac{2N_{TT}}{N} - \frac{N_{FT}}{N} = \frac{\lambda a(a-1)}{1+\lambda a}$$

It is easy to prove that $c(\Omega) = c(\frac{\lambda a(a-1)}{1+\lambda a}) = \frac{\lambda a(a-1)}{1+\lambda a} = \Omega(a, \lambda)$. In this way we discover the fixed part of Sugeno complement that depends on the fuzzy truth value "a" and of the parameter λ. When $\lambda > 0$, we have that $\Omega < 0$.

Note that:

(1) Ω is the density of the worlds where p is fixed.
(2) $\Omega < 0$ is the density of worlds where the false value is fixed. Thus the density $a = \dfrac{N_p}{N}$ is the true value of p with an external parameter λ.

Discussion:

(1) The conditions $a = 0$ and $a = 1$ are the boundary conditions. In this case $\Omega = 0$ and all the worlds for the NOT operator change in a coherent way with the logic value of the proposition p. In this case, we have a total synergy among the worlds for which the different operators NOT act in the same time.
(2) For $\lambda > 0$ then $\Omega < 0$. When "a" changes from 0 to 1 then Ω decreases from 0 to a minimum value Ωmin , and it then increases from a negative value to the value zero.
(3) For $-1 < \lambda < 0$ then $\Omega > 0$. When "a" changes from 0 to 1 then Ω increases from 0 to its maximum value of Ωmax , after which it decreases to the value zero.

We note that the equilibrium point for which $c(a,\lambda) = a$ is the minimum value for Ω.when $\lambda > 0$. It is the maximum value for Ω when $-1 < \lambda < 0$.

We also note that when $\Omega < 0$, the worlds where NOT is inactive (the logic value of p does not change) are the worlds where p = "FALSE". When $\Omega > 0$, the worlds where NOT is inactive are the worlds where p ="TRUE".

When $\lambda > 0$, $\Omega < 0$, the number of worlds where p = "TRUE" increases from 0. The number of worlds where p = FALSE and NOT is inactive increases until the equilibrium point is reached.

The reverse occurs when $-1 < \lambda < 0$, $\Omega > 0$. That is, the number of worlds where p = "TRUE" increases from 0 until the equilibrium point is reached while "NOT is inactive" is increased. From the equilibrium point, the number of worlds decreases where p = "TRUE" and "NOT is inactive" for the value a = 1 where NOT is active.

Note that,

$$c\,[\,\Omega(a\,,\lambda\,)\,] = \Omega(a\,,\lambda\,).$$

The function $\Omega(a, \lambda)$ is the equilibrium function or invariant for the complement operation.

6.5.3 Action of the Agent and Operator Γ

Given the expression $F = \neg\,p \wedge p$, we assume that a fundamental symmetry exists for this expression. The same proposition p is also present in two different positions within the expression F. When this global symmetry is written as $F = \neg\,p \wedge p$, we have that the law of contradiction is true. But when the symmetry is broken:

$$F = \neg\,\Gamma p \wedge p \tag{6.14}$$

In Figure 6.6 we show the action of the operator Γ:

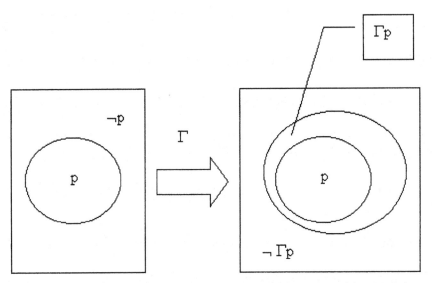

Figure 6.6. Showing the set of worlds where Γp is true. In this case it is larger than the set where p is true.

Here Γp is a new variable for which the equation (6.15) is not always false.

De Morgan law:

In the classical logic we have:

$$\neg\,(\,p \wedge q\,) = (\neg\,p \vee \neg\,q\,) \text{ and } \neg\,(\,p \vee q\,) = (\neg\,p \wedge \neg\,q\,)$$

It is shown that in fuzzy logic

$$\neg\,(\,p \wedge q\,) \to \neg\,\Gamma\,(\,p \wedge q\,) = \Gamma\,\neg\,(\,p \wedge q\,) = \Gamma\,(\neg\,p \vee \neg\,q\,)$$

is true. In fuzzy logic when the same logic expression is rewritten the logic expression is different and not equivalent to itself.

Example:

In the classical logic $p \wedge q$ may be rewritten as follows:

$$p \wedge q \equiv (p \vee q) \wedge (\neg p \vee q) \wedge (p \vee \neg q)$$

In fuzzy logic we know that when we rewrite the expression $p \wedge q$, we generate a new expression:

$(p \vee q) \wedge (\neg \Gamma p \vee q) \wedge (p \vee \neg \Gamma q)$ which in general differs from $p \wedge q$.

Example:

Consider the set of the worlds S:

$S = \{w_1, w_2, w_3, w_4, w_5\}$

Suppose that evaluation of p is given as:

$V(p) = \{$True, False, True, True, False$\}$

And,

$V(\neg p) = \{$False, True, False, False, True$\}$

We then have $\mu_p = \dfrac{3}{5}$ and $\mu_{\neg p} = \dfrac{2}{5}$. However for $F = p \wedge \neg p$ we have:

$\mu_{p \wedge \neg p} = 0$ so $\mu_{p \wedge \neg p} \neq \min(\mu_p, \mu_{\neg p}) = \dfrac{2}{5}$

Returning to fuzzy rules we use the Γ operator as follows:

$V(\neg \Gamma p) = \{$True, False, False, True, False$\}$

Here the logic value of the world 1 is permuted with the logic value of the world 2 and the logic value of the world 5 is permuted with the logic value of the world 4. We then have:

$$\mu_{p \wedge \neg \Gamma p} = \frac{2}{5} \text{ and } \mu_{p \wedge \neg \Gamma p} = \min (\mu_p , \mu_{\neg \Gamma p}) = \frac{2}{5}$$

6.5.4 Normal Forms, which Occur When We Break the Symmetry in the Law of Contradiction

For the previous considerations and for the linguistic "AND" operator we have:

$$\Gamma p \wedge \Gamma q \equiv (\Gamma p \vee \Gamma q) \wedge (\neg \Gamma p \vee \Gamma q) \wedge (\Gamma p \vee \neg \Gamma q) \quad (6.15)$$

In (6.15) we have a fundamental symmetry. For each variable the operator Γ is the same in every occurrence. But when we break the previous symmetry and we use the operator Γ_1 in the logic complement "$\neg$" occurrence and Γ_2 in all the other occurrences, the previous identity is not true.

So the (6.15) for $\Gamma_1 \neq \Gamma_2$ becomes:

$$\Gamma_2 p \wedge \Gamma_2 q \neq (\Gamma_2 p \vee \Gamma_2 q) \wedge (\neg \Gamma_1 p \vee \Gamma_2 q) \wedge (\Gamma_2 p \vee \neg \Gamma_1 q) \quad (6.16)$$

The (6.16) can be proved as follows:

(i) DNF: $\Gamma p \wedge \Gamma q \rightarrow \Gamma_2 p \wedge \Gamma_2 q$

(ii) CNF : $(\Gamma p \vee \Gamma q) \wedge (\neg \Gamma p \vee \Gamma q) \wedge (\Gamma p \vee \neg \Gamma q)$

$\rightarrow (\Gamma_2 p \vee \Gamma_2 q) \wedge (\neg \Gamma_1 p \vee \Gamma_2 q) \wedge (\Gamma_2 p \vee \neg \Gamma_1 q)$

$$= (\Gamma_2 p \vee \Gamma_2 q) \wedge ([\neg \Gamma_1 \Gamma_2^{-1}] \Gamma_2 p \vee \Gamma_2 q) \wedge (\Gamma_2 p \vee [\neg \Gamma_1 \Gamma_2^{-1}] \Gamma_2 q)$$

$$= (\Gamma_2 p \vee \Gamma_2 q) \wedge (c \Gamma_2 p \vee \Gamma_2 q) \wedge (\Gamma_2 p \vee c \Gamma_2 q)$$

Where,

DNF is the disjunctive normal form and

CNF is the conjunctive normal form (see Türkşen, 1986).[26]

For the expression (i), we can write:

$$\Gamma_2 p \wedge \Gamma_2 q = [(p \wedge \neg SB_2(p)) \vee (\neg p \wedge SB_2(p))] \wedge [(q \wedge \neg SB_2(q)) \vee (\neg q \wedge SB_2(q))]$$

For the expression (ii), we can write:

$$(\Gamma_2 p \vee \Gamma_2 q) \wedge (\neg \Gamma_1 p \vee \Gamma_2 q) \wedge (\Gamma_2 p \vee \neg \Gamma_1 q)$$
$$= \Gamma_2 p \wedge \Gamma_2 q \vee [\neg \Gamma_1 q \wedge \Gamma_2 q \vee \neg \Gamma_1 p \wedge \Gamma_2 p] \tag{6.17}$$

Note that when $\Gamma_1 q = \Gamma_2 q$ and $\Gamma_1 p = \Gamma_2 p$, the two normal forms are equal and the law of contradiction is true.

For the expression (6.17), when $\Gamma_1 p \neq \Gamma_2 p$ and $\Gamma_1 q \neq \Gamma_2 q$, we have:

$$[\neg \Gamma_1 q \wedge \Gamma_2 q \vee \neg \Gamma_1 p \wedge \Gamma_2 p]$$
$$= [\neg q \wedge \neg SB_1(q) \wedge SB_2(q)] \vee [q \wedge SB_1(q) \wedge \neg SB_2(q)] \vee$$
$$\vee [\neg p \wedge \neg SB_1(p) \wedge SB_2(p)] \vee [p \wedge SB_1(p) \wedge \neg SB_2(p)]$$

For the expressions, (i) and (ii), when $\Gamma_1 p \neq \Gamma_2 p$ and $\Gamma_1 q \neq \Gamma_2 q$, we have:

$\mu\,(\,\Gamma_2\,p \wedge \Gamma_2\,q\,) \le \mu\,(\,\Gamma_2\,p \wedge \Gamma_2\,q \vee [\,\neg\,\Gamma_1\,q \wedge \Gamma_2\,q \vee \neg\,\Gamma_1\,p \wedge \Gamma_2\,p\,]\,)$

That is DNF is included in CNF. Thus when the different symmetry break operators, $\Gamma_1 \ne \Gamma_2$, are applicable, then we get DNF$\ne$CNF and in particular DNF$\subseteq$CNF.

For the linguistic operation "OR" we have:

(i) CNF: $\Gamma p \vee \Gamma q \;\to\; \Gamma_2\,p \vee \Gamma_2\,q$

(ii) DNF: $(\Gamma p \wedge \Gamma q\,) \vee (\neg\Gamma p \wedge \Gamma q\,) \vee (\Gamma p \wedge \neg\Gamma q\,) \;\to\; (\Gamma_2 p \wedge \Gamma_2 q) \vee (\neg\Gamma_1 p \wedge \Gamma_2 q\,) \vee (\Gamma_2 p \wedge \neg\Gamma_1 q)$

Thus, DNF$\subseteq$CNF. is applicable.

6.5.5 Quantum Logic [17]

The formalism of quantum mechanics, built mainly to make experimental forecasting within the reality of microscopic physics receives important contribution from quantum logic. The idea of a "logic of quantum mechanics" or quantum logic was originally suggested by Birkhoff and Von Neumann [39] in their 1936 pioneering paper. Here we present only a brief summary of the fundamentals of quantum logic. Let us start from an n-tuple of state wave functions:

$(\psi_1, \psi_2, \ .., \psi_n\,)$

which are common eigenfunctions of a set of operators $A_1, A_2, \ , A_q$ (associated with suitable observables). Because the n-tuple is the same for every observable, the observables are called compatible observables. Mathematically this happens if and only if the associated operators commute. An example of a pair of compatible observables is energy and momentum of a free particle. The set of all

functions generated by a linear combination of the functions belonging to the n-tuple constitute a universal set of functions. On the
power set of this universal set, we can define a Boolean algebra in
the usual way. This algebra is a mathematical structure for classical
logic, with the traditional operations of conjunction, disjunction
and negation. In this way, as Birkhoff and Von Neumann pointed
out, we can associate a classical logic lattice with every set of compatible observables. The situation is very different, however, when
we deal with pairs of incompatible observables, such as position
and momentum of a free particle. In this case the position operator
is associated with a particular n-tuple of eigenfunctions, whereas
the momentum operator is associated with a different n-tuple of eigenfunctions. As it is well known, the two operators do not commute. As a consequence, only one universal set of functions is not
sufficient. We are forced to introduce multiple universal sets (in
this case two).

Of course, classical logic is not applicable in such situation. A new
logic is needed, called quantum logic. One particular difference of
quantum logic with respect to classical logic is its lack of distributive law. This means that in general:

$$p \wedge (q \vee r) \neq (p \wedge q) \vee (p \wedge r)$$
$$p \vee (q \wedge r) \neq (p \vee q) \wedge (p \vee r)$$

In quantum logic the contradiction law is true but is not true for the
distributive property. In classical logic we have:

$$p \wedge (q \vee r) = (p \wedge q) \vee (p \wedge r)$$

When we break the symmetry and the operator Γ is applied to every
variable, the distributive law still holds after a break in symmetry in
every world.

$$\Gamma p \wedge (\Gamma q \vee \Gamma r) = (\Gamma p \wedge \Gamma q) \vee (\Gamma p \wedge \Gamma r)$$

However when we have a break of symmetry of the second order, and:

$$(\Gamma p \wedge \Gamma q) \vee (\Gamma p \wedge \Gamma r) \rightarrow (\Gamma_1 p \wedge \Gamma_2 q) \vee (\Gamma_1 p \wedge \Gamma_2 r)$$
$$\Gamma p \wedge (\Gamma q \vee \Gamma r) \rightarrow \Gamma_2 p \wedge (\Gamma_2 q \vee \Gamma_2 r)$$

We then have:

$$(\Gamma_1 p \wedge \Gamma_2 q) \vee (\Gamma_1 p \wedge \Gamma_2 r) \neq \Gamma_2 p \wedge (\Gamma_2 q \vee \Gamma_2 r)$$

The distributive property is no longer true. The contradiction law:

$$\neg \Gamma_1 p \wedge \Gamma_1 p = \neg \Gamma_2 p \wedge \Gamma_2 p = \text{FALSE}$$

then applies.

Example:

Consider,

$$S = \{w_1, w_2, w_3, w_4, w_5\}$$

Suppose the valuation of p is given as:

$V(p) = \{\text{True, False, True, True, True}\}$ and $V(\Gamma_2 p) = \{\text{False, True, True, True, False}\}$

$V(\Gamma_1 p) = \{\text{True, False, True, True, True}\}$

The values of q and r are:

$V(q) = \{\text{True, True, True, False, False}\}$,

$V(\Gamma_2 q) = \{\text{True, True, True, False, True}\}$

$V(r) = \{\text{True, False, True, False, True}\}$,

$V(\Gamma_2 r) = \{$False, True, True, True, False $\}$

We then have:

$V[\Gamma_2 p \wedge (\Gamma_2 q \vee \Gamma_2 r)] = \{$False, True, True, True, False $\}$

$V[(\Gamma_1 p \wedge \Gamma_2 q) \vee (\Gamma_1 p \wedge \Gamma_2 r)] = \{$True, False, True, True, True $\}$

The membership of

$\Gamma_2 p \wedge (\Gamma_2 q \vee \Gamma_2 r)$ is 2/5

and the membership of

$[(\Gamma_1 p \wedge \Gamma_2 q) \vee (\Gamma_1 p \wedge \Gamma_2 r)$ is 1/5.

Remark:

In the classical logic we can represent an expression in different forms which are equivalent to one another. In the fuzzy logic when we change the form of an expression we obtain a new expression which is similar to the first, but not equivalent. This is the basic difference between the normal logic forms and fuzzy logic forms.

6.6 Invariance in Fuzzy Logic

In classical logic we study tautology expressions which are always true for every logic value of the variables. Every logic model is based on tautology. When we study fuzzy logic we have logic values in a set of real numbers between 0 and 1. In this case the traditional tautology becomes invariant and we can try to create a fuzzy logic model similar to the classical modal logic by invariant properties.

6.6.1 Invariance in One Variable

We know that in classical logic we have the simple tautology:

$$T = p \lor \neg p \text{ that is always true for every logic value of p} \qquad (6.18)$$

When we evaluate the previous expression by the worlds and introduce an external agent that uses the operator Γ to change the logic value for p , we obtain the fuzzy expression of T or TF:

$$T F = p \lor \Gamma \neg p = p \lor \neg \Gamma p \qquad (6.19)$$

We introduce a new variable or Γp in (6.19), and this is not formally equal to (6.18). We can then assume also false value when the variable Γp is different from the variable p. For equation (6.19) we can write, for every world a relationship:

$$T F = \Gamma p \to p \quad \text{ that is false when } \Gamma p \text{ is true and p is false.}$$

Such a tautology in the classical logic is however not a tautology in fuzzy logic. In fact using the Zadeh operation we have:

$$\mu_{p \text{ or } cp} = \max(\mu_p, 1 - \mu_p) \le 1$$

To obtain an invariant within the fuzzy logic, we change the variables in the expression (6.18) as follows:

$$p \to p \lor \neg \Gamma p$$

As the expression (6.18) is a tautology, this is always true for every substitution of the variable. So we have that:

$$T = (p \lor \neg \Gamma p) \lor \neg (p \lor \neg \Gamma p) \quad \text{ is always true}$$

Consequently T can be divided into two parts. One part is the expression (6.19) and the other part is the compensation part. In conclusion for the DeMorgan rule we have:

$$T = (p \vee \neg \Gamma p) \vee \neg p \wedge \Gamma p \qquad\qquad (6.20)$$

Example:

When we use the following evaluation:

$V (p) = (\text{True} , \text{True} , \text{False} , \text{False} , \text{False}) , V (\neg \Gamma p) = (\text{True} , \text{True} , \text{True} , \text{False} , \text{False})$

$V (\Gamma p) = (\text{False} , \text{False} , \text{False} , \text{True} , \text{True}) , V (\neg p) = (\text{False} , \text{False} , \text{True} , \text{True} , \text{True})$

Obtaining the relationship:

$V (p \vee \neg \Gamma p) = (\text{True} , \text{True} , \text{True} , \text{False} , \text{False})$ and

$V (\neg p \wedge \Gamma p) = (\text{False} , \text{False} , \text{False} , \text{True} , \text{True})$

Returning to the original fuzzy logic we obtain:

$$\max [\mu_p, (1 - \mu_p)] + \min [(1 - \mu_p) , \mu_p] = 1$$

Where $\max [\mu_p, (1 - \mu_p)]$ is associated with the expression $V (p \vee \neg \Gamma p)$ and $\min [(1 - \mu_p) , \mu_p]$ with $V (\neg p \wedge \Gamma p)$.

The term $\min [(1 - \mu_p) , \mu_p]$ is the Compensation Term. We can extend the same process to other tautologies and we can find new invariants in fuzzy logic.

6.6.2 Modus Ponens (Inferential Rule)

In the classical logic we can write the Modus Ponens as follows:

$$p \wedge q = p \wedge (p \rightarrow q) \qquad (6.21)$$

The left part of the expression (6.21) is a static part that is true only when p and q are true. The right part of (6.21) is the dynamical part. When p and ($p \rightarrow q$) are true, for the left part we have that q is also true. So we move the true value from p to q by the implication rule $p \rightarrow q$.

Because $(p \rightarrow q) = \neg p \vee q$, we can write the previous expression using the operator Γ as:

$\neg \Gamma p \vee q$ and the Modus ponens are:

$$p \wedge q = p \wedge (\Gamma p \rightarrow q) \quad \text{when} \quad \Gamma p = p \qquad (6.22)$$

$$p \wedge q \neq p \wedge (\Gamma p \rightarrow q) \quad \text{when} \quad \Gamma p = \neg p$$

In the second equation in (6.22) we have the fuzzy modus ponens. In fact when p is true and ($\Gamma p \rightarrow q$) is true q is false. Considering the implication of $\Gamma p \rightarrow q$ as the axiomatic form, when p is true it is not always true that q is true. There exist worlds where we cannot use the inferential process. To obtain a compensation term for (6.22) we substitute p in (6.21) with the variable $p \wedge \Gamma p$ which contains the gamma operator. Remembering that the (6.21) is a tautology and so is true in every case for every variable, we obtain:

$$(p \wedge \Gamma p) \wedge q \equiv (p \wedge \Gamma p) \wedge [\neg(p \wedge \Gamma p) \vee q] \qquad (6.23)$$

which is always true. The previous expression can be written as:

$$(p \wedge \Gamma p) \wedge q \equiv (p \wedge \Gamma p) \wedge (\neg p \vee \neg \Gamma p \vee q)$$

or,

$$(p \wedge \Gamma p) \wedge q \equiv (p \wedge \Gamma p \wedge \neg p) \vee (p \wedge \Gamma p) \wedge (\neg \Gamma p \vee q)$$

or,

$$(p \wedge \Gamma p) \wedge q \equiv (p \wedge \Gamma p) \wedge (\Gamma p \to q) \qquad (6.24)$$

or finally as the expression:

$$(p \wedge q) \wedge \Gamma p \equiv [p \wedge (\Gamma p \to q)] \wedge \Gamma p \qquad (6.25)$$

where the expression (6.22) is present. In the expression (6.24) we compensate equation (6.22) by substituting p with the expression $(p \wedge \Gamma p)$. When $(p \wedge \Gamma p)$ is true p must be true and also Γp and so $\Gamma p = p$. We limit our study only to the subset of worlds where $\Gamma p = p$ and we eliminate the worlds where $\Gamma p = p$ is false. In (6.25), we have the same expression as in (6.22) with the limitation for the case where $\Gamma p = p$.

It is noted that given a classical tautology we can rewrite the tautology using a fuzzy logic operation. When the fuzzy logic operation is introduced, the tautology loses its property and is then not always true. Using suitable substitutions in the classical tautology, we can check where and why the fuzzy image of the tautology is not always true. At the same time we can give the fuzzy logic expression which compensates the fuzzy image of the tautology in such a way so as to restart the original property: that is the relationship is true in every case. We can build fuzzy invariant expressions that are analogous to the fuzzy logic expressions of classical tautology. A bridge between fuzzy logic and classical logic can be established using Γ operators.

6.7 Conclusion

In this chapter we have studied the case where the adaptive agent at different orders cannot compensate for all of the different rules that reach one given context. When many different sets of rules are present in the same context, an ambiguous situation exists because we must have one and only one set of rules for every context.

Here the adaptive agent is able to formalize uncertainty and generate compensation by using the concept of world. In one world we can have only one set of rules that can be evaluated in a logic way. The definition of the concept of worlds is not sufficient to explain the meaning of uncertainty. In the classical modal logic the introduction of the concept of a possible world is not related to an uncertainty process. The world is in fact the instrument used to determine if a proposition is true or false. In the meta-theory evaluation in one world is extended to all other worlds (generalization). The generalization is a non standard process which generates confusion and uncertainty. For example consider the case with two worlds and one proposition p. When in one world we define the proposition p and in the other world we define $\neg$ p, the logic evaluation is true in one world and false in the other world. Because we have a superposition of p and $\neg$ p, confusion or uncertainty exist. However when we generalize the evaluation from one world to another, the same proposition p is assumed to be true or false in both worlds for the evaluation in the first world. But it is possible that p is present in one world and $\neg$ p is present in the other world. In this case the generalization is false.

In this chapter we have shown that the average value of the true value for p is associated with the membership function obtained by using the fuzzy set theory. The adaptive agent has a fractional number that measures the incoherence within the context. When the degree of incoherence or membership function is equal to one, the process of compensation has eliminated every conflict and consequently the proposition p is true or false for every world. Using the image of the possible world and the generalization we can activate a logic computation by the classical operations AND; OR, NOT. With different configurations of the true values in the different worlds and using the classical logic operation in every world it is possible to prove that we can reproduce all the fuzzy logic operations. To move from one fuzzy operation to another we introduce the Semantic Event in which special agents can change the logic value in the worlds. The membership function in the set of the worlds is a semantic measure which is isomorphic to the probability measure. The main difference between the ordinary events and the semantic events is that the first cannot change the logic value of the proposition while the second can change the logic value.

The adaptive agent within uncertainty can find compensatory fuzzy logic terms which can eliminate uncertainty and generate logic invariants for which the membership value is always equal to one. In conclusion if the adaptive agent cannot compensate the system so as to have a coherent context, it can measure how far it is removed from the coherence and use a fuzzy logic computation.

Bibliography

1. Chellas B.F., *Modal Logic: An Introduction*. Cambridge University Press, 1980.

2. Dempster A.P., "Upper and Lower Probabilities induced by Multi valued Mapping", *Annals of Mathematical Sciences*, vol. 38, pp. 325-339, 1967.

3. Dubois D. and Parade H., *Possibility Theory*. Plemun Press, 1988.

4. Harmanec D., Klir G.J. and Resconi G., "On Modal Logic Interpretation of Dempster-Shefer Theory of Evidence", *Int. J. Intelligent Systems*, vol.9, pp. 941-951, 1994.

5. Hughes G.E. and Cresswell M.J., *An Introduction to Modal Logic*. Methuen, 1968.

6. Klir G.J. and Harmanec D., "On Modal Logic Interpretation of Possibility Theory", *Int. J. of Uncertainty, Fuzziness and Knowledge-Based Systems*, vol.2, pp. 237-245, 1994.

7. Klir G.J. and Yuan B., *Fuzzy Sets and Fuzzy Logic*. Prentice Hall, 1995.

8. Murai T., Nakata M. and Shimbo M., "Ambiguity, Inconsistency, and Possible-Worlds: A New Logical Approach", *Proceedings of the Ninth Conference on Intelligence Technologies in Human-Related Sciences*, Leòn, Spain, pp. 177-184, 1996.

9. Murai T., Kanemitsu H. and Shimbo M., "Fuzzy Sets and Binary-Proximity-based Rough Sets", *Information Science*, vol.104, pp. 49-80, 1998.

10. Pawlak Z., *Rough Sets*, Kluwer, Netherland, 1991.

11. Resconi G., Klir G.J. and St. Clair U., "Hierarchical Uncertainty Metatheory Based Upon Modal Logic", *Int. J. of General Systems*, vol. 21, pp. 23-50, 1992.

12. Resconi G., Klir G.J., St. Clair U. and Harmanec D., "On the Integration of Uncertainty Theories", *Int. J. of Uncertainty, Fuzziness, and Knowledge-Based Systems*, vol. 1, pp. 1-18, 1993.

13. Resconi G. and Rovetta R., *Fuzzy Sets and Evidence Theory in a Metatheory Based Upon Modal Logic*, Quaderni del Seminario Matematico di Brescia, n.5, 1993.

14. Resconi G., Klir G.J., Harmanec D. and St. Clair U., "Interpretations of Various Uncertainty Theories Using Modals of Modal Logics: A Summary, Fuzzy Sets and Systems", vol. 80, pp. 7-14, 1996.

15. Resconi G. and Murai T., "Field Theory and Modal Logic by Semantic Field to Make Uncertainty Emerge from Information", *Int.J.General System*, 2000.

16. Resconi G. and Türkşen I.B., "Canonical forms of fuzzy truths by meta-theory based upon modal logic", *Information Sciences*, vol. 131, pp.157 – 194, 2001.

17. Resconi G., Klir G.J. and Pessa E., "Conceptual foundation of quantum mechanics the role of evidence theory, quantum sets, and modal logic", *Int. J. of Modern Physics C*, vol.10, no.1, pp.29-62, 1999.

18. Shafer G., *A Mathematical Theory of Evidence*, Princeton University Press, 1976.

19. Sowa J.F., *Knowledge Representation: Logical, Philisophical, and Computational Foundation.* PWS Publishing Co., Pacific Grove, CA., 1999

20. Türkşen I.B., "Upper and Lower Set Formulas: Restriction and Modification of Dempster-Pawlak Formalism", *Special Issue of Int. J. of Applied Mathematics and Computer Science*, vol.12, no.3, 2002.

21. Türkşen I.B., "Theories of Set and Logic with Crisp and Fuzzy Information Granules", *J. of Advanced Computational Intelligence*, vol.3, no.4, pp.264-273, 1999.

22. Türkşen I.B., Computing with Descriptive and Veristic Words, Proceedings of NAFIPS'99 (Special Invited Presentation), New York, pp.13-17, June 10-12, 1999.

23. Türkşen I.B., "Fuzzy Normal Forms", FSS, pp. 253-266, 1994.

24. Türkşen I.B., "Interval-Valued Fuzzy Sets and 'Compensatory AND'", FSS, pp. 295-307, 1994.

25. Türkşen I.B., "Non-Specificity and Interval-Valued Fuzzy Sets", FSS, pp. 87-100, 1995.

26. Türkşen I.B., "Interval-Valued Fuzzy Sets Based of Normal Forms", FSS, pp. 191-210, 1986.

27. Türkşen I.B., Kandel A. and Zhang Y.Q., "Normal Forms of 'Fuzzy Middle' and 'Fuzzy Contradiction'", *IEEE-SMC*, vol.29 Paert B (Cybernetics), no.2, pp.237-253, 1999.

28. Türkşen I.B., "Computing with Descriptive and Veristic Words: Knowledge Representation and Reasoning", in: P.P. Wang (ed.)

Computing with Words, Wiley Series on Intelligent Systems, J. Wiley&Sons, New York, pp.297-328, 2001.

29. Zadeh L.A., "Fuzzy Sets, Information and Control", vol.8, pp. 338-353. 1965.

30. Zadeh L.A., "A Simple View on the Dempster-Shafer Theory of Evidence and Its Implication for the Rule of Combination", Al Magazine, vol. 7, pp.85-90, 1986.

31. Zadeh L.A., "Fuzzy Logic = Computing With Words", *IEEE-Trans on Fuzzy Systems*, vol.42, pp.103-111, 1996.

32. Zadeh L.A., "Computing With Perceptions", Keynote Address, *IEEE- Fuzzy Theory Conference*, San Antonio, May 7-10, 2000.

33. Zimmerman H.J. and Zysno P., "Latent Connectives in Human Decision Making", FSS, pp. 37-51, 1980.

34. Roberto Serra, Massimo Andreatta, Giovanni Canarini, Mario Compiani, *Introduction to the Physics of Complex systems / The mesoscopic approach to fluctuations, non linearity and self – organization*, Pergamon Press, 1986.

35. Resconi G., Murai T. and Shimbo M., "Field theory and modal logic by semantic fields to make uncertainty emerge from information", *Int. J General Systems*, vol.29, no.5, pp.737 782, 2000.

36. Muirhead H., *The Physics of Elementary Particle*, Pergamon Press, 1965.

37. Milutin Blagojevic, *Gravitation and Gauge Symmetries*, IOP Publishing, Bristol, 2002.

38. Klir G.J. and Yuan B., *Fuzzy Sets and Fuzzy Logic (Theory and Application)*, Prentice Hall Upper Saddle River, New Jersey, 07458, 1995.

39. Birkhoff G. and von Neumann J., Ann. Math. vol.37, pp.823, 1936.

Chapter Seven

The Hierarchical Structure of Adaptive Agents

7.1 System Coupling and Adaptive Agents

An important definition found in the System Theory (A. Wayne Wymore, 1993) [6] is the System Coupling:

Given many systems or agents at the first order, the conjunctive system is illustrated in Figure 7.1.

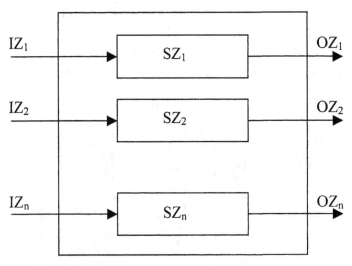

Figure 7.1. A conjunctive system or agent.

Every agent or system in the Figure 7.1 is defined in the usual way:

$$Z = (\ SZ\ ,\ IZ\ ,\ OZ\ ,\ NZ\ ,\ RZ\ )$$

Where,

SZ is a finite set which is not empty
IZ is a finite set
OZ is a finite set
NZ: $SZ \times IZ \rightarrow SZ$ or including time we have
NZ: $SZ(t) \times IZ \rightarrow SZ(t+1)$
RZ: $SZ \rightarrow OZ$

Every agent can be represented by these triples as shown in Table 7.1.

Table 7.1. Ontological representation of the simple system or agent.

System Z	Source	Propriety	Value
NZ	State at time t $\in$ SZ	Value Input $\in$ IZ	State at time t+1$\in$ SZ
RZ	State $\in$ SZ	Output function	Output $\in$ OZ

The set of triples gives the ONTOLOGY of the system Z. For the conjunctive system or agent we have the set of triples.

Table 7.2. Ontological representation of the conjunctive system or agent.

	Source	Property	Value
NZ_1	$SZ_1(t)$	$IZ_1(t)$	$SZ_1(t+1)$
RZ_1	$SZ_1(t)$	RZ_1	$OZ_1(t)$
NZ_2	$SZ_2(t)$	$IZ_2(t)$	$SZ_2(t+1)$
RZ_2	$SZ_2(t)$	RZ_2	$OZ_2(t)$
....	....	....	....
NZ_n	$SZ_n(t)$	$IZ_n(t)$	$SZ_n(t+1)$
RZ_n	$SZ_n(t)$	RZ_n	$OZ_n(t)$

An example of two conjunctive systems:

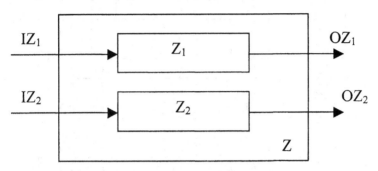

Figure 7.2. A conjunctive system with two systems Z_1 and Z_2.

The conjunctive system Z can be represented by the transformation of couples of states in Z_1 and in Z_2. The read out function connects a couple of states with a couple of output. In the Table 7.3 we give the ontology of the conjunctive system Z when Z_1 and Z_2 are:

$Z_1 = (S Z_1 , I Z_1 , O Z_1 , N Z_1 , R Z_1)$
$S Z_{1-} = \{1,2\}$
$I Z_1 = \{3,4\}$
$O Z_1 = \{5,6 \}$
$N Z_1 = \{(1,3),1) , ((1,4),2), ((2,3),1), ((2,4),2) \}$
$R Z_1 = \{(1,5) , (2,6) \}$

and,

$Z_2 = (S Z_2 , I Z_2 , O Z_2 , N Z_2 , R Z_2)$
$S Z_2 = \{1,2\}$
$I Z_2 = \{5,6\}$
$O Z_2 = \{3,4\}$
$N Z_2 = \{(1,5),1) , ((1,6),2), ((2,5),1), ((2,6),2) \}$
$R Z_2 = \{(1,3) , (2,4) \}$

Table 7.3. A conjunctive system with two systems Z_1 and Z_2.

	Source	Property	Value
NZ	(1,1)	(3,5)	(1,1)
NZ	(1,2)	(3,5)	(1,1)
NZ	(2,1)	(3,5)	(1,1)
NZ	(2,2)	(3,5)	(1,1)
NZ	(1,1)	(3,6)	(1,2)
NZ	(1,2)	(3,6)	(1,2)
NZ	(2,1)	(3,6)	(1,2)
NZ	(2,2)	(3,6)	(1,2)
NZ	(1,1)	(4,5)	(2,1)
NZ	(1,2)	(4,5)	(2,1)
NZ	(2,1)	(4,5)	(2,1)
NZ	(2,2)	(4,5)	(2,1)
NZ	(1,1)	(4,6)	(2,2)
NZ	(1,2)	(4,6)	(2,2)
NZ	(2,1)	(4,6)	(2,2)
NZ	(2,2)	(4,6)	(2,2)
	(1,1)	RZ	(5,3)
	(1,2)	RZ	(5,4)
	(2,1)	RZ	(6,3)
	(2,2)	RZ	(6,4)

7.1.1 Connection in Cascade as Adaptive Agent

When communication exists between Z_1 and Z_2 where for example the output of Z_1 using the channel of communication R is sent to the input of Z_1, we have the well known connection between systems which is known as cascade. In Figure 7.3 we show the cascade connection.

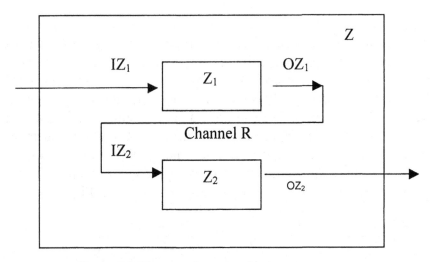

Figure 7.3. The cascade connection between systems.

The cascade can also be represented by the set of triples in Table 7.4 by which we denote the ontology image of the cascade.

Table 7.4. Ontology image of the cascade connection.

	Source	Property	Value
NZ_1	$SZ_1(t)$	$IZ_1(t)$	$SZ_1(t+1)$
RZ_1	$SZ_1(t)$	RZ_1	$OZ_1(t)$
NZ_2	$SZ_2(t)$	$OZ_1(t)$	$SZ_2(t+1)$
RZ_2	$SZ_2(t)$	RZ_2	$OZ_2(t)$

We see in Table 7.4 that within the ontology we have this equation or condition:

$$OZ_1 = IZ_2 \qquad\qquad (7.1)$$

Between Z_1 and Z_2 we have the channel R. There are several possible different situations. The channel R transforms OZ_1 so the agent Z_2 receives an erroneous value of OZ_1. On the channel R passes a message OZ_1 that is incompatible with the input IZ_2.

We remember that every second order agent generates an image of the rules in the first context that is projected into the second context. The image enters the second context and is then compared with the rules in the second context. This can also generate conflicts or inconsistencies between the image and the local rules. In the cascade connection we can have a similar situation. In the first context we have the inputs IZ_1 and rules that transform the input value. By the system Z_1 we have an image of IZ_1 denoted as the output OZ_1. The image must then be compared with the input value IZ_2 and the transformations in the second context. In Figure 7.4 we show the context image of the cascade:

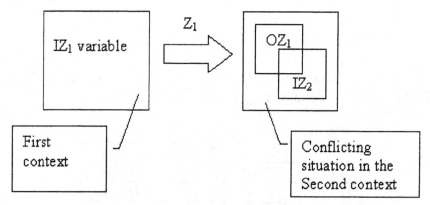

Figure 7.4. When Z_1 projects an image OZ_1 into the second context. We must then compare OZ_1 with IZ_2, where $IZ_2 = R \, OZ_1$ and R is the channel of communication. We remark that we have the same structure for the action as the second order adaptive agent.

The cascade can be modeled as an action of the second order of an adaptive agent.

The adaptive agent compensates every incongruence generated by the system Z_1 and in the channel R .We remember from Chapter 2 that Table 7.5 gives the second order action of an adaptive agent.

Table 7.5. Ontology of the second order action

	Source	Property	Value
Statement 1	IN_1	X_1	IN_2
Statement 2	IN_1	P_1	OUT_1
Statement 3	IN_2	P_2	OUT_2
Statement 4	OUT_1	X_2	OUT_2

In the cascade system we have the ontology of action as shown in Table 7.6.

Table 7.6. Ontology of the second order action for a cascade system.

	Source	Property	Value
Statement 1	IZ_1	X_1	IZ_1
Statement 2	IZ_1	Z_1	OZ_1
Statement 3	IZ_1	Z_1	$OZ_1 = OUT_2$
Statement 4	OZ_1	R	$IZ_2 = OUT_2$

It can be seen in the cascade that the coherence condition is given by the equation:

$$IZ_2 = OZ_1 \text{ or } IZ_2 = R\,OZ_1$$

where R is the channel of communication.

Example:

Given the two systems Z_1 and Z_2. The system Z_1 is:

$Z_1 = (S\,Z_1 , I\,Z_1 , O\,Z_1 , N\,Z_1 , R\,Z_1)$
$S\,Z_{1\text{-}} = \{1,2\}$
$I\,Z_1 = \{3,4\}$
$O\,Z_1 = \{5,6\}$
$N\,Z_1 = \{(1,3),1) , ((1,4),2), ((2,3),1), ((2,4),2) \}$
$R\,Z_1 = \{(1,5) , (2,6) \}$

and the system Z_2 is:

$Z_2 = (S\, Z_2\, ,\, I\, Z_2\, ,\, O\, Z_2\, ,\, N\, Z_2\, ,\, R\, Z_2\,)$
$S\, Z_2 = \{1,2\}$
$I\, Z_2 = \{5,6\}$
$O\, Z_2 = \{5,6\}$
$N\, Z_2 = \{(1,3),1)\, ,\, ((1,4),2),\, ((2,3),1),\, ((2,4),2)\, \}$
$R\, Z_2 = \{(1,5)\, ,\, (2,6)\, \}$

In the Table 7.7 we show the coherent connection in cascade of the systems Z_1 and Z_2. We have:

$RZ_1\, (SZ_1) = OZ_1 = R\, OZ_1.$

Table 7.7. Coherent cascade connection between the systems Z_1 and Z_2.

	Source	Property	Value
	3	X_1	3
	4	X_1	4
NZ_1	1	3	1
NZ_1	1	4	2
NZ_1	2	3	1
NZ_1	2	4	2
	1	RZ_1	5
	2	RZ_1	6
NZ_1	1	3	1
NZ_1	1	4	2
NZ_1	2	3	1
NZ_1	2	4	2
	1	RZ_1	5
	2	RZ_1	6
	5	R	5
	6	R	6

The ontology of the cascade system Z is given by the Table 7.7. In the Table 7.7 the next state function NZ transforms the couple of states in Z_1 and Z_2 to another couple of states. The read out function RZ gives the output value OZ_2 when we know the couple of states in Z_1 and in Z_2.

We remark that OZ_1 and IZ_1 are not present in Z. In fact OZ_1 and IZ_1 are considered as internal output and internal input. In the Table 7.8 we assume that coherence between the internal output and input exists. The system Z is always in the meta-stable situation because the internal communication between the two systems Z_1 and Z_2 can change to become a no faithful communication. In this situation we must start the second order system into which we introduce the adaptive agent that can restore the communication channel and join the two systems or agents in one system or agent Z.

In Table 7.8 we show by an asterisk the incoherent case for the cascade when we introduce the internal output OZ_1 and the internal input IZ_2.

Table 7.8. Among all possible transformations of the states we show by an asterisk the incoherent case in the cascade connection between the systems Z_1 and Z_2.

	Source	Property	Value
NZ	(1,1)	(3,5)	(1,1)
NZ	(1,2)	(3,5)	(1,1)
NZ	(2,1)	(3,5) *	(1,1)
NZ	(2,2)	(3,5) *	(1,1)
NZ	(1,1)	(3,6) *	(1,2)
NZ	(1,2)	(3,6) *	(1,2)
NZ	(2,1)	(3,6)	(1,2)
NZ	(2,2)	(3,6)	(1,2)
	(1,1)	(4,5)	(2,1)
	(1,2)	(4,5)	(2,1)
	(2,1)	(4,5) *	(2,1)
	(2,2)	(4,5) *	(2,1)
	(1,1)	(4,6) *	(2,2)
	(1,2)	(4,6) *	(2,2)
	(2,1)	(4,6)	(2,2)
	(2,2)	(4,6)	(2,2)
	(1,1)	RZ	(5,3)
	(1,2)	RZ	(5,4)
	(2,1)	RZ	(6,3)
	(2,2)	RZ	(6,4)

When we eliminate the internal output and the internal input and assume that all the incoherent cases are eliminated, we obtain the possible transformations shown in the Table 7.9.

Every cascade for the coherent situation transforms only a couple of state subsets. An adaptive agent must ensure that no anomalous or incoherent transformation can enter the cascade system.

Table 7.9. System Z that is the cascade of the systems Z_1 and Z_2.

	Source	Property	Value
NZ	$(1,1)$	3	$(1,1)$
NZ	$(1,2)$	3	$(1,1)$
NZ	$(2,1)$	3	$(1,2)$
NZ	$(2,2)$	3	$(1,1)$
NZ	$(1,1)$	4	$(2,1)$
NZ	$(1,2)$	4	$(1,1)$
NZ	$(2,1)$	4	$(2,2)$
NZ	$(2,2)$	4	$(2,2)$
	$(1,1)$	RZ	5
	$(1,2)$	RZ	6
	$(2,1)$	RZ	5
	$(2,2)$	RZ	6

In Figure 7.5 we show the second order Adaptive Agent which controls the coherence of the internal communication between OZ_1 and IZ_2.

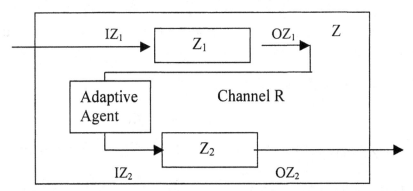

Figure 7.5. Cascade of two systems with a second order adaptive agent which controls the internal coherence for $OZ_1 = IZ_2$.

7.2   The Feedback as a Second Order System

Feedback is a special case of cascade where the external input and output are eliminated. We have only internal output and input. We introduce, as in the cascade, the adaptive agent into the internal connection between output and input

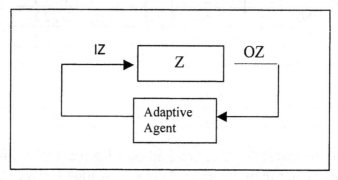

Figure 7.6. Feedback as a special case of the cascade with a second order adaptive agent that controls the internal coherence where OZ = IZ.

In the previous chapters we have defined the discrete system by the model:

$$Z = (\,SZ\,,IZ\,,OZ\,,NZ\,,RZ\,)$$

Another function is added in the feedback. Given the function:

$$RZ : SZ \to OZ \quad \text{and}$$

$$NZ : SZ \; x \; IZ \to SZ$$

we have,

$$OZ = RZ\,[\,NZ\,(\,SZ\,,IZ\,)\,] \tag{7.2}$$

For the feedback we have:

Table 7.10. Feedback ontology.

	Source	Property	Value
NZ	SZ(t)	IZ = OZ	SZ(t+1)
RZ	SZ(t+1)	RZ	OZ = IZ

As OZ = IZ for the coherent feedback, we have:

$$IZ = RZ\,[\,NZ\,(\,SZ\,,\,IZ\,)\,] \tag{7.3}$$

Given that the input IZ at time t for the system Z and with feedback it will have another value of IZ at time t+1. The feedback system FZ is formally given as:

$$FZ = (\,SZ\,,\,IZ\,,\,F\,) \tag{7.4}$$

Where F is the expression (7.3).

$$F\,(\,SZ\,,\,IZ\,) = RZ\,[\,NZ\,(\,SZ\,,\,IZ\,)\,]$$

Example:

The system Z with the next state function NZ and the read out function RZ is given in Table 7.11.

Table 7.11. System Z.

	Source	Property	Value
NZ	1	A	2
NZ	1	B	3
NZ	1	C	2
NZ	1	D	3
NZ	2	A	4
NZ	2	B	4
NZ	2	C	4
NZ	2	D	4
NZ	3	A	4
NZ	3	B	4
NZ	3	C	4
NZ	3	D	4
NZ	4	A	2
NZ	4	B	2
NZ	4	C	2
NZ	4	D	2
	1	RZ	B
	2	RZ	A
	3	RZ	D
	4	RZ	C

The feedback state and the source values and the input values at time t and time t+1 are given in Table 7.12.

Table 7.12. System Z and the function IZ = RZ [NZ (SZ , IZ)].

	Source	Property	Value
RZ NZ	A	1	A
RZ NZ	A	2	C
RZ NZ	A	3	C
RZ NZ	A	4	A
RZ NZ	B	1	D
RZ NZ	B	2	C
RZ NZ	B	3	C
RZ NZ	B	4	A
RZ NZ	C	1	A
RZ NZ	C	2	C
RZ NZ	C	3	C
RZ NZ	C	4	A
RZ NZ	D	1	D
RZ NZ	D	2	C
RZ NZ	D	3	C
RZ NZ	D	4	A

The function IZ = RZ [NZ (SZ , IZ)] given in Table 7.12 can be represented also in a graphic way in Figure 7.7.

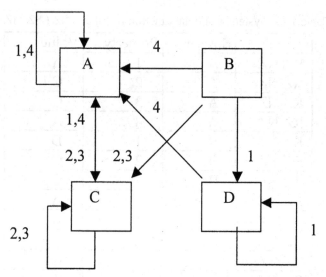

Figure 7.7. The connection between the inputs for different values of the system given in Table 7.10 for different states. The graph is obtained by using the relationship IZ = RZ [NZ (SZ , IZ)] under the control of the second order adaptive agent.

The transition from one input to another input is controlled by the states of the system Z. When the state is stable and the input does not change, there is a stable feedback. In the previous example the system Z has no stable states and consequently the feedback has not a fixed point and the input changes in a continuous manner.

A recursive process:

$$IZ(t+1) = RZ [NZ (SZ , IZ(t))]$$ (7.5)

can represent all feedbacks.

Every feedback is a deterministic recursive process controlled by the states of the system. From the initial input we can determine the next input. The input in the feedback is not free as in a usual system but changes with time under the control of the states. When we have a recursive process with a fixed point or stable point in the feedback it is stable. Any random process changes the initial value of the input but it cannot destroy the recursive process which is completely deterministic. When the feedback is subject to an external random disturbance, the feedback is out of the control. However when the random disturbance ceases, the feedback activates the deterministic recursive process and the basic underlying behavior returns.

For example in Figure 7.8 we have two different modes of behavior. One of these is a movement which returns to a stable point, the other is a cyclic movement.

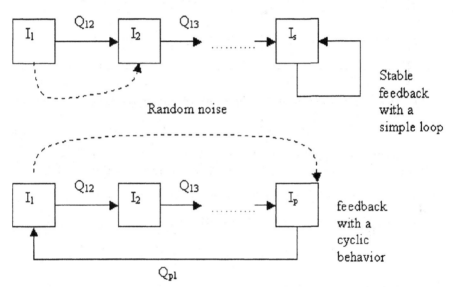

Figure 7.8. Showing two different patterns of behavior for input values I under the control of the states Q. The effect of the random external noise is to change the initial value of feedback. It cannot change the recursive process and the resulting type of the behavior that is a stable or oscillatory output.

In the feedback the set of inputs is the same as the set of outputs. Between the input values and the output values there is a one time step delay.

7.3 General Feedback with the System Z_2

Every feedback can be obtained by the use of two different systems Z_1 and Z_2 as follows:

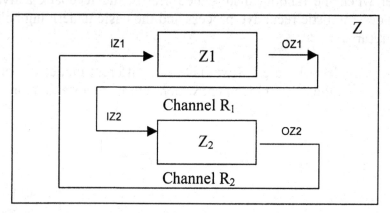

Figure 7.9. Extension of the simple feedback system by the introduction of the system Z_2.

The feedback in Figure 7.9 can be considered as a simple feedback system having two systems in cascade as shown in Figure 7.10.

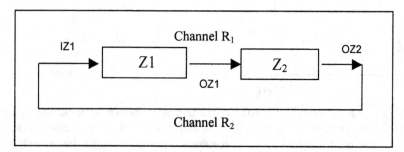

Figure 7.10. Similar to the simple feedback system shown in Figure 7.9 but having two systems Z_1 and Z_2 in cascade.

Given the two systems:

$Z_1 = (SZ_1 , IZ_1 , OZ_1 , NZ_1 , RZ_1)$

And,

$$Z_2 = (SZ_2 , IZ_2 , OZ_2 , NZ_2 , RZ_2)$$

When cascaded as shown in Figure 7.10, the two systems can be fused into one system Z as:

$$Z = [SZ, IZ_1 , OZ_2 , NZ , RZ_2]$$

Where $SZ = (SZ_1, SZ_2)$ is the Cartesian product of the states of Z_1 and Z_2 . Here NZ is the next state function defined as:

$$NZ = NZ_2 [RZ_1 NZ_1 (IZ_1 , SZ_1) , SZ_2]$$

The read out value of Z is as follows:

$$OZ_2 = RZ_2 NZ_2 [RZ_1 NZ_1 (IZ_1 , SZ_1) , SZ_2]$$

For the feedback we have the recursive process:

$$IZ_1(t+1) = RZ_2 NZ_2 [RZ_1 NZ_1 (IZ_1(t) , SZ_1) , SZ_2] \qquad (7.6)$$

The recursive process of the feedback is under the control of two states one of which is SZ_1 and the other state is SZ_2.

Example:

For the two systems Z_1 and Z_2 in the Table 7.13

Table 7.13. Ontology of the systems Z_1 and Z_2.

	Source	Property	Value
NZ_1	1	5	1
NZ_1	1	6	2
NZ	2	5	1
NZ_1	2	6	2
	1	RZ1	3
	2	RZ1	4
NZ_2	1	3	1
NZ_2	1	4	2
NZ_2	2	3	1
NZ_2	2	4	2
	1	RZ2	5
	2	RZ2	6

The system Z is given in Table 7.14 as follows:

Table 7.14. The system Z which fuses the two cascaded systems Z_1 and Z_2.

	Source (SZ_1, SZ_2)	Property IZ_1	Value
$NZ = NZ_2 [RZ_1 NZ_1 (IZ_1 , SZ_1) , SZ_2]$	(1,1)	5	(1,1)
	(1,2)	5	(1,1)
	(2,1)	5	(1,1)
	(2,2)	5	(1,1)
	(1,1)	6	(2,2)
	(1,2)	6	(2,2)
	(2,1)	6	(2,2)
	(2,2)	6	(2,2)
	1	RZ_2	5
	2	RZ_2	6

For the system Z in Table 7.14 and equation (7.6) the feedback function is:

$$F = RZ_2 NZ_2 [RZ_1 NZ_1 (IZ_1(t) , SZ_1) , SZ_2] = IZ_1(t+1) \qquad (7.7)$$

And the relationship between the inputs $IZ_1(t)$ (source) and the inputs $IZ_1(t+1)$ (value) is given in Table 7.15.

Table 7.15. Feedback function F in (7.7) for the system in Table 7.14.

	Source	Property	Value
	5	F	5
	6	F	6

The feedback is given by the recursive function:

$$IZ_1(t+1) = RZ_2\, NZ_2\, [\, RZ_1\, NZ_1\, (\, IZ_1(t)\, ,\, SZ_1\,)\, ,\, SZ_2\,].$$

For every value of the state SZ_2 we have a simple feedback, we can choose both the system Z_2 and the state SZ_2 so as *to choose the desired change of input IZ_1.* In many cases the second system Z_2 can ensure stable behavior for IZ_1. We show in Figure 7.11 a possible change in the behavior of the input IZ_1 by the use of Z_2. The left diagram in Figure 7.11 is the function F in Table 7.14. But with a change of the system Z_2 in Table 7.13, we can obtain another function F as shown in the right part of Figure 7.11. The system Z_2 can so change the behavior of the feedback from a stable behavior into a cycling behavior.

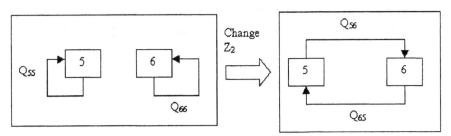

Figure 7.11. Showing one possible change of the behavior of the input IZ_1 by using Z_2.

The feedback with two systems or agents can be represented as a communication between a master agent Z_1 and a slave agent Z_2. This is illustrated in Figure 7.12.

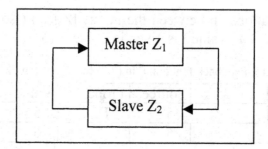

Figure 7.12. The cascaded systems Z_1 and Z_2 are represented as being a communication between two agents. The former is the master agent and the latter is the slave.

7.3.1 An Example Comparing Feedback with the Second Order Action

Given the algebraic equation:

$$AX^2 + B X + C = 0 \qquad\qquad (7.8)$$

Where A, B, C are parameters. We consider the quadratic form Y = $AX^2 + B X + C$ as a system where Y is the output , X is the input and (A , B , C) are the states of the system for different values of A, B and C. We construct a feedback system that changes the input X in such a way as to reach a stable condition where Y is either zero or near zero for a given approximation. (Refer Figure 7.13).

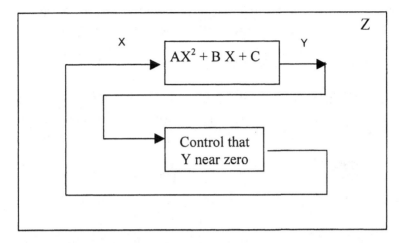

Figure 7.13. For the system $Y = AX^2 + B X + C$ the control feedback changes the input value x until y is near zero with a desired approximation.

The solution of the equation (7.8) can be found using another level of abstraction when we substitute a feedback process which changes the input dynamically using a second order action. We create two different contexts. Using one context we locate all possible transformations of translations of the input value. We substitute a change of the control system input using an unknown translation.

$$X' = X + \eta$$

Using the other context we locate the symbolic algebraic forms $P (X) = AX^2 + B X + C$ and possible transformations. We know that every variable in input X has been associated with symbolic algebraic form. We have:

$$X \rightarrow P (X)$$

$$X + \eta \rightarrow A(X + \eta)^2 + B (X + \eta) + C$$

In the second context using all possible transformations of the symbolic forms, we need the transformations of the type:

$$AX^2 + B X + C \rightarrow A X^2 + \beta$$

Where the linear term is eliminated. Now we show in Figure 7.14 the action at the second order and the conflict which arises.

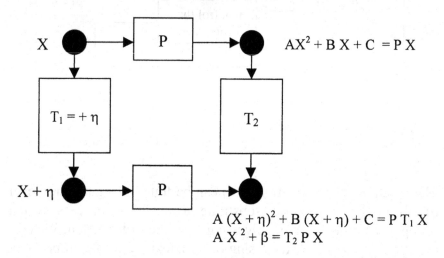

$$A (X + \eta)^2 + B (X + \eta) + C = P T_1 X$$
$$A X^2 + \beta = T_2 P X$$

Figure 7.14. The first context with the translation transformations is on the left side. On the right side the second context transformation T_2 is shown. It eliminates the linear term in the quadratic form. The transformation from the left context into the right context is a second order action.

The action in the Figure 7.14 is not coherent. This is because in general we have $A X^2 + \beta$ which is different from $A (X + \eta)^2 + B (X + \eta) + C$. Using the compensation agent we try to solve the coherent equation:

$$A (X + \eta)^2 + B (X + \eta) + C = A X^2 + \beta \qquad (7.9)$$

From Equation (7.9) we have:

$$2 A \eta + B = 0 \text{ and } \beta = A\eta^2 + B \eta + C$$

Resulting in:

$\beta = (- B^2 + 4 A C) / 4 A$

Since $A X^2 + \beta = 0$ can be solved immediately, we have:

$X = \pm (\sqrt{\beta}) / 2 A$

The solution X on the left part of (7.9) is substituted:

$\pm (\sqrt{\beta}) / 2 A + \eta = - B / 2A \pm (\sqrt{\beta}) / 2 A$

which is the solution of the equation:

$A (X + \eta)^2 + B (X + \eta) + C = 0$

When we substitute $Z = X + \eta$ we have

$Z = - B / 2A \pm (\sqrt{\beta}) / 2 A$ which is the solution of the equation:

$A Z^2 + B Z + C = 0$

Using the coherent condition at the second order, the adaptive agent finds the general solution of the equation (7.8). Using feedback only an approximate solution can be found.

7.3.2 Second Order Action to Improve The Convergence of The Feedback

Given a particular feedback we know that the input value IZ_1 changes by a recursive process as follows:

$IZ(t+1) = F(SZ , IZ(t))$ (7.10)

Where,

$F(SZ, IZ(t)) = RZ [NZ (SZ, IZ(t))]$ or

$F(SZ, IZ(t)) = RZ_2 NZ_2 [RZ_1 NZ_1 (IZ_1(t), SZ_1), SZ_2]$.

The equation (7.10) can also be written as:

$IZ(t+1) = F(SZ, IZ(t)) + IZ(t) - IZ(t)$

Or,

$IZ(t+1) = G(SZ, IZ(t)) + IZ(t)$

where,

$$G(SZ, IZ(t)) = F(SZ, IZ(t)) - IZ(t) = IZ(t+1) - IZ(t) \qquad (7.11)$$

For the *stable situation* we have,

$[IZ_1(t+1) - IZ_1(n)] = 0$ and $G(IZ_1(t), SZ_1) = 0$

When we take a recursive process in *the first context* it is always convergent as:

$$IZ_1(t+1)] = \begin{cases} -\dfrac{IZ_1(t)^2}{1+IZ_1(t)^2} + IZ_1(t) \text{ if } IZ_1(0) > 0 \\[4mm] \dfrac{IZ_1(t)^2}{1+IZ_1(t)^2} + IZ_1(t) \text{ if } IZ_1(0) < 0 \end{cases} \qquad (7.12)$$

When we move from the *first context* to the *second context* using function G in (7.11) then,

$$IZ_1(t+1)] = \gamma \frac{G[IZ_1(t)]^2}{1+G[IZ_1(t)]^2} + IZ_1(t) \qquad (7.13)$$

The new recursive process (7.12) is stable when G = 0. That is (7.11) is stable when the original feedback is stable.

Because $\dfrac{G[IZ_1(n)]^2}{1+G[IZ_1(n)]^2} \geq 0$ when $IZ_1(0) < IZ_1^*$ and $G(IZ_1^*) = 0$.

We must put $\gamma = -1$ and in the other cases we put $\gamma = 1$. Because at the beginning we do not know the solution IZ_1^*, we observe the feedback for the first few steps and if the feedback diverges we change the sign of the parameter γ.

When we move from one context to another context we assume that the variable IZ_1 in 7.12 changes in $G(IZ_1)$ with the same invariant rule.

7.4 Second Order Action which Transforms a Feedback in One Context into a Feedback in Another Context

In chapter 7.3 we show that it is possible to control one feedback using a system Z_2 as shown in Figure 7.12. The system Z_2 changes the behaviour of the feedback from unstable to a stable fixed point. Using the second order action we can change the feedback by the operators P_1 and P_2 into the second order action that we repeat in Table 7.16.

Table 7.16. Action of the second order as given in chapter 2.

Statement	Resource	Property	Value
S_1	IN_1	X_1	IN_2
S_2	IN_1	P_1	OUT_1
S_3	IN_2	P_2	OUT_2
S_4	OUT_1	X_2	OUT_2

Where $IN_1 = IZ_1(t)$, $IN_2 = IZ_1(t+1)$, $OUT_1 = IZ_2(t)$, $OUT_2 = IZ_2(t+1)$ and $X_1 = Z_1$ and $X_2 = Z_2$ we note that Z_2 is the image of the system Z_1.

In Figure 7.15 we show the two contexts with two feedbacks.

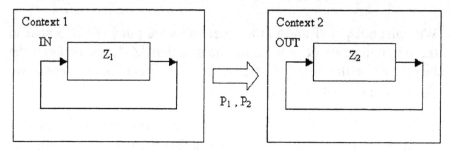

Figure 7.15. By using the operators P_1 and P_2 we project the feedback from the first context into the second context.

We remember the internal relation:

$$P_2 X_1 IZ_1(t) = X_2 P_1 IZ_1(t) = OUT_2 \tag{7.14}$$

When we write the simple statement in a formal mathematical way:

Feedback first context: $IZ_1(t) \rightarrow IZ_1(t+1)$

Also,

Feedback second context: $P_1 IZ_1(t) \rightarrow P_2 IZ_1(t+1)$

In Figure 7.16 we show the action which transforms one feedback in one context into a different feedback in another context.

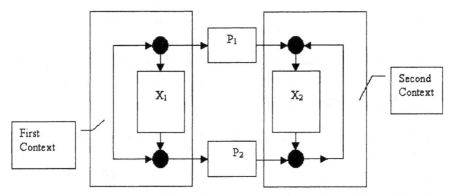

Figure 7.16. In the First Context we have the First Feedback. The Second Feedback on the right is the Projection of the First Context by the operators P_1 and P_2.

Example:

In Figure 7.7 we have shown the transition from one input to another by use of feedback. Now by using second order action we can change this transition diagram in different ways. This was discussed in chapter 2. We give an example of this in Figure 7.17.

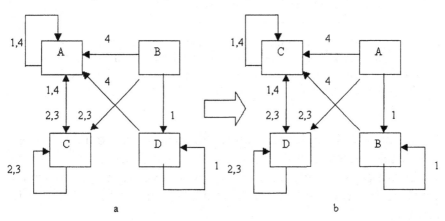

Figure 7.17. Showing the transformation of the Transition of the Input in one Feedback into another transition in a Second Context Feedback.

7.5 Hyper-feedback (Hyper-cycle)

In the book "The Hyper-cycle" written by M. Eigen and P. Schuster [1], the authors introduce the Hyper-cycle into the biological system and we synthesize as follows.

We consider a sequence of reactions in which the last product formed is identical with a reactant of the first step (autocatalytic system). The catalyst chain of reactions is a cycle (feedback) shown in Figure 7.18.

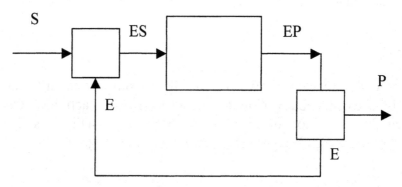

Figure 7.18. Illustrating the biological cycle (autocatalytic system). or biological feedback that transforms the input chemical molecule S (substrate) into the chemical product P. The production is mediated by the chemical molecule E (Enzyme) which does not change in the loop but is used to obtain the product P.

M. Eigen and P. Schuster described other more complex cycles which they denoted as Hyper-cycles for which each part of these hyper-cycles is a cycle. A catalytic hyper-cycle is a system which connects autocatalytic or self – replicate unities through a cyclic linkage.

An example of the biological hyper-cycle or hyper-feedback is given by the Bethe – Weizsacker hyper-cycle which obtains glucose from the light of the sun, the water and CO_2 by a cycle of autolocatylic reactions. Another important Hyper-cycle is the Krebs cycle which transforms glucose into water , CO_2 and energy by a cycle of autolocatylic reactions. The two cycles are complementary to one another. The Bethe – Weizsacker cycle is used by plants to generate glucose, while glucose in the Krebs cycle is used by plants and animals to obtain energy useful for their life.

M. Eigen and P. Schuster described some models of the Hyper-cycles. In our book we suggest a new approach to the hyper-cycle using the action of agents at the second or higher order as we show in Figure 7.19 where we describe a feedback whose elements are feedbacks. For the definition of M. Eigen and P. Schuster in Figure 7.19 we have a Hyper-feedback.

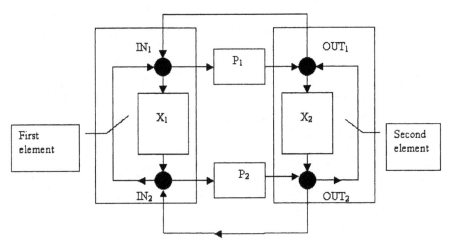

Figure 7.19. Describing the feedback result of the second order action which is the same as the element on which the action acts. Because the elements of the actions are feedbacks, we have feedback of feedbacks or hyper-feedback.

Here the feedback in the first context changes in time. One feedback goes to another feedback in a dynamical way. In the Hyper – feedback we have two scales of the time. The time with short unity changes the input values inside the single feedback. The time with

higher unity changes the feedback itself. In Figure 7.20 we show
the diagram of diagrams in the Hyper – feedback at the second order, where G_1, G_2, G_3, G_4 are particular elements of hyper-
feedback.

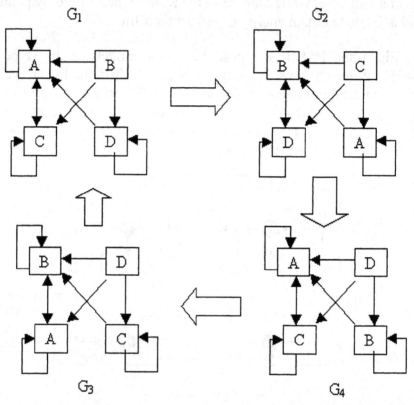

Figure 7.20. Showing the transition from one diagram G_α to another in
the Hyper-feedback . Within the small diagram the transitions are at the
time t. Each of the diagrams G_1, G_2 , G_3 , G_4 represents a later time T
where $T > t$.

7.6 Third Order Action which Transforms a Second Order Hyper – Feedback

In Figure 7.21 one Hyper –feedback in one context is transformed into another in another context. The change can be used to change the hyper-diagram as we show in Figure 7.21, into other inputs at different times.

In Figure 7.21 with the operators A_1, A_2, A_3, A_4 described in chapter 2, we realize an action for an agent at the third order. In this action a hyper –diagram as in Figure 7.20 is transformed into another hyper-diagram as we show in Figure 7.21. In this situation we have three different times. One is the time t for the transition in the simple diagrams G , the other is the time T for the transition within the super-diagram that changes the diagrams G and the other **T** is the time that changes one super-diagram into another super diagram. We generate by the different orders of actions a nested set of diagrams with an ordered set of times.

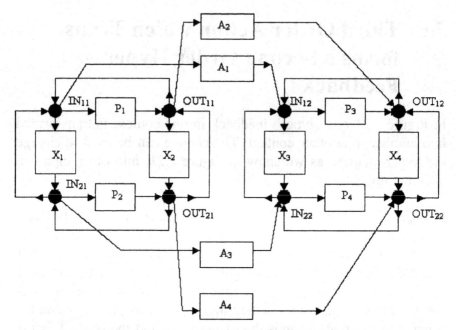

Figure 7.21. Transformation of the Hyper –feedback from one context to another.

In Figure 7.22 below we show the transformation from one super-diagram into another.

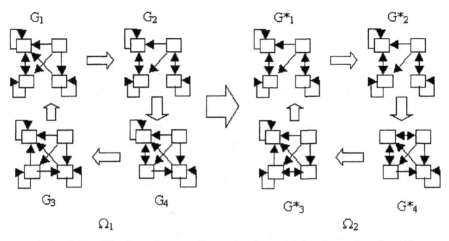

Figure 7.22. We show the transformation by a third order action from the
super diagram Ω_1 into the super diagram Ω_2.

We can use the same process to generate Hyper- feedback at the
third , fourth or greater orders.

7.7 Conclusion

In this chapter we have studied two cascaded systems using
Adaptive Agents of the second order which control the
communication between the two systems. A special case of cascade
is that of feedback. This has been studied exhaustivily and an
explanation of the transition diagram for the inputs is given. The
feedabck and a possible abstract extension by second order action
is also given. After the study of simple feedback, Hyper-feedbacks
at different orders have been considered.

Bibliography

1. Eigen M. and Schuster P., "The Hypercycle C: The realistic hypercycle". Naturwissenschaften, vol.65, pp.341-369, 1978.

2. Eigen M., "Selforganization of matter and the evolution of macromolecules". Naturwissenschaften, vol.58, pp.465-523, 1971.

3. Eigen M. and Schuster P., "The Hypercycle A: Emergence of the hypercycle", Naturwissenschaften, vol.64, pp.541-565, 1977.

4. Eigen M. and Schuster P., "The Hypercycle B: The abstract hypercycle", Naturwissenschaften, vol.65, pp.7-41, 1978.

5. Famulok M., Nowick J. and Rebek Jr. J., "Self- Replicating Systems". Act. Chim.Scand, vol.46, no.4, pp.315-324, 1992.

6. Wymore A.W., *Model-Based Systems Engineering*. CRC Press, 1993.

Chapter Eight

The Adaptive Field in Logical Conceptual Space

8.1 Introduction

With the definition of the Adaptive Field we may assign a particular degree of significance to any possible world in modal logic. In this way, our attention can be focused on more significant worlds or on the comparison between two different worlds. The information is one of the main sources able to show the logic structure of information. A guide is obtained from this on the use of information. It is also able to discover and measure the degree of uncertainty contained. The space of worlds is thus useful in its ability to divide information into the important parts. An accessibility relation exists between the most significant worlds with respect to the least significant worlds. Logical models of information are possible using the adaptive fields.

The mathematical description of fields and Adaptive Fields gives us a new possibility to obtain models of logic which depend on the contained information.

8.2 Adaptive Field

Resconi, Klir, Harmanec, and St.Clair have recently developed in a series of papers [4, 6, 15, 16, 17, 18] a new hierarchical uncertainty metatheory based upon the use of modal logic. This new approach

307

enables us to collect various different uncertainty theories which include Fuzzy Set Theory [22], the Dempster-Shafer Theory of Evidence [19, 23], (where the theory of evidence is an extension of the probability calculus when elements are sets.) and Possibility Theory in one metatheory.

The most difficult task in the use of the metatheory is to define the possible worlds and the accessibility relationship between possible worlds. In order to overcome the difficulties, we now propose the concept of "Degree of Significance", which is the support for the Adaptive Field in the space of possible worlds. By the Adaptive Field and Degree of Significance we define the accessibility relations. The Adaptive Field is the mathematical support for *introspective* information of a human being and it uses the "Field Theory" as a tool. The meta-theory uses as Inferential Process a tool to describe the degree of uncertainty in the information and also to discover the logic structure within the data.

The type of Kripke Model used for modal logic is dependent on the structure of the Adaptive Field. The logic is not fixed but is both flexible and dynamic depending on the particular field in a space time reference. A completely sure event is described by the use of only one world as shown in chapter 6. When uncertainty arises, the world will be divided into many possible worlds. In this way we can generate the required space of worlds to obtain the Belief and Plausibility measures required by the Dempster-Shafer Theory [19].

We use the variation of the Adaptive Field as the accessibility relation. One world is accessible to another world when its Adaptive Field is higher than the other. Adaptive Field information establishes the level of significance for any world. Because the Adaptive Field is generated by the available information, we can obtain a logic image of the data without special pre-processing. Any Adaptive Field is a virtual field that can be implemented by a physical field such as an electromagnetic field, a chemical concentration field or any other field.

We know that a system with an infinite number of degrees is a field [10]. A field associates one value with each point of the space. The number of points in a geometric space is infinite; consequently the number of values of the field is infinite. When we consider the points of the space as inputs of a system and the associated value as the output of the system, the number of degrees of freedom of the system is equal to the number of the points of the space, so it is infinite.

The equation of motion for a field as an *individual entity* (*the field equations*), can be obtained by the principle of the least action. Given the generic position x in the space, the field function is F(x). The equation of motion of the field is obtained by this minimal condition:

$$\delta A = \delta \int L \, dx = 0 \tag{8.1}$$

Here δ is the symbol of variation, dx = dtdxdydz and L is the Lagrangian density in the space time reference. We know that the minimal action condition gives the equation of motion for the field F(x). The equation of the field is the Euler-Lagrange equation:

$$[\partial L / \partial F - (\partial / \partial x_\lambda)(\partial L / \partial(\partial F / \partial x_\lambda)) = 0 \tag{8.2}$$

For example, when the Lagrangian density is:

$$L = a (\partial F / \partial x_\lambda)^2 + b F$$

the dynamics of the field is given by the equation:

$$(\partial_{xx} + \partial_{yy} + \partial_{zz} - c^{-1} \partial_{tt} - b/a) F = 0$$

All the transformations of the space time for which δA is invariant:

$$\delta A = \delta \int L \ dx = \delta \int L' \ dx' \qquad (8.3)$$

lead to the appearance of certain conservation laws for the fields associated with the Lagrangian, see Noether's theorem [10]. In chapter 2 we have shown that the fundamental aim of physics is to find symmetry within the universe. This means to find invariants for every possible context or situation in the physical space time domain. For the (8.3) at every change of the space time domain, the variation of the action A never changes. Noether proved that this invariance includes all the classical principles of invariance for energy, momentum and others. Noether simplified the representation of the fundamental symmetry (invariance) in the physical laws.

A given derivative operator in one context or reference changes its form and its value when we move to another context. The derivative in the new context loses its former properties. To restore these properties we introduce an Adaptive Field into the new context. The compensating derivative is denoted as the covariant derivative. The covariant derivative D_s is related to the previous derivative:

$$\frac{\partial F_k(y^j)}{\partial y^j}$$ of the field $F_k(y^j)$ as follows;

$$D_s F'_p = \frac{\partial F_k(y^j)}{\partial y^j} \frac{\partial y^j}{\partial x^s} \frac{\partial y^k}{\partial x^p} \qquad (8.4)$$

Here x^k are the coordinates in the new reference or context and F'_p is the field in the new context. We note that the Lagrangian form is invariant when we pass from the derivative to the covariant derivative. The equations:

$$L = L \, [\, F_k (\, y_\lambda), \, \frac{\partial F_k (y_\lambda)}{\partial y_\lambda} \,] \text{ and } L = L \, [\, F'_k (\, x_\lambda), \, D_\lambda \, F'_k (\, x_\lambda) \,] \quad (8.5)$$

are the Lagrangian density in the old reference and in the new one with the same form. When $T = \dfrac{\partial y^j}{\partial x^i}$, we show the covariant derivative in Figure 8.1 (For more details on this diagram and theory related to this diagram see the GSLT theory [14] in chapter 3).

In (8.5) we remark that the Lagrangian depends on the simple variable $F_k (\, y_\lambda)$, and on the variable $\dfrac{\partial F_k (y_\lambda)}{\partial y_\lambda}$. When we change the reference in the space- time or other spaces, we move from one context to another and the derivative changes. To restore a derivative similar to the original derivative in the first context (symmetry in field theory), we must compensate the derivative (covariant derivative) in the second context as shown in Figure 8.1.

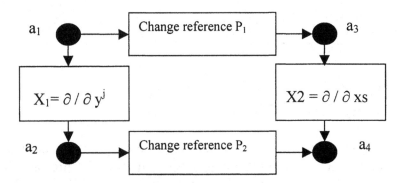

Figure 8.1. Second order action that changes the simple derivative on the left into the covariant derivative D_s on the right.

In Figure 8.1 we have:

$$a_1 = F_k \, (y_j) = F_k.$$

This is because F_k is a vector and when we change the reference it changes as follows:

$$a_3 = F'_p = F_k \frac{\partial y^k}{\partial x^p}$$

Here $a_2 = \dfrac{\partial F_k}{\partial y^j}$ is a tensor and it becomes $\dfrac{\partial F_k}{\partial y^j} \dfrac{\partial y^j}{\partial x^s} \dfrac{\partial y^k}{\partial x^p}$.

We then have:

$$a_4 = [\frac{\partial F_k}{\partial y^j} \frac{\partial y^j}{\partial x^s} \frac{\partial y^k}{\partial x^p} , \; \frac{\partial F_k}{\partial y^j} \frac{\partial y^j}{\partial x^s} \frac{\partial y^k}{\partial x^p} + F_k \frac{\partial^2 y^k}{\partial x^s \partial x^p}]$$

The second expression in a_4 is the derivative of $F_k \dfrac{\partial y^k}{\partial x^p}$ in the second context. This becomes:

$$\frac{\partial F'_p}{\partial x^s} = \frac{\partial F_k}{\partial x^s} \frac{\partial y^k}{\partial x^p} + F_k \frac{\partial^2 y^k}{\partial x^s \partial x^p}$$

$$= \frac{\partial F_k}{\partial y^j} \frac{\partial y^j}{\partial x^s} \frac{\partial y^k}{\partial x^p} + F_k \frac{\partial^2 y^k}{\partial x^s \partial x^p}$$

Using the same definition of the derivative in the first context and second context, we have a conflicting situation. In this case the internal coherence equation is not true as outlined in Chapter 2. That is:

$$X_2 P_1 \neq P_2 X_1$$

When the definition of the derivative is changed in the new reference in this way:

$$D_s F'_p = \frac{\partial F'_p}{\partial x^s} - F_k \frac{\partial^2 y^k}{\partial x^s \partial x^p}$$

$$= \frac{\partial F_k}{\partial y^j} \frac{\partial y^j}{\partial x^s} \frac{\partial y^k}{\partial x^p}$$

In this case in a_4 we have only one value for the two paths:

$$a_4 = [\frac{\partial F_k}{\partial y^j} \frac{\partial y^j}{\partial x^s} \frac{\partial y^k}{\partial x^p}, \frac{\partial F_k}{\partial y^j} \frac{\partial y^j}{\partial x^s} \frac{\partial y^k}{\partial x^p}]$$

and $X_2 P_1 = P_2 X_1$.

The new derivative $\frac{\partial F_k}{\partial y^j} \frac{\partial y^j}{\partial x^s} \frac{\partial y^k}{\partial x^p}$ is not equal to the derivative $\frac{\partial F_p}{\partial x^s}$ but is similar when we change the reference. This is very important because all the relations that include the derivative $\frac{\partial F_p}{\partial x^s}$ are invariant for the change of the reference when we use the derivative $\frac{\partial F_k}{\partial y^j} \frac{\partial y^j}{\partial x^s} \frac{\partial y^k}{\partial x^p}$. We satisfy the fundamental symmetric property of physics underlined in chapter 2.

The field $F_k \frac{\partial^2 y^k}{\partial x^s \partial x^p}$ adapts the derivative of the second context so as to produce a coherence condition. So we call:

$$H = F_k \frac{\partial^2 y^k}{\partial x^s \partial x^p} \text{ the } \textit{adaptive field} \tag{8.6}$$

The space where we define the fields is the space of the worlds. In chapter 6 we have defined the possible world and we have given a brief introduction to modal logic. In this chapter we define a space where at any point we define a world. The space where we operate the change of reference is not a physical space but is a logic space of the worlds. For modal logic, at every point of this conceptual space of the worlds we can know if one sentence is true or false. In chapter 5 we have discussed and used the space of the general co-ordinates. In this chapter we use a similar reference with a different meaning of the points.

8.3   Degrees of Significance as Adaptive Field and Accessibility Relations

In this section we show a way of generating an accessibility relation between possible worlds using degrees of significance. In order to generate the relationship we need some further information on the possible world. This information forms a context and each piece of information, connected with a relevant possible world, constitutes a local world. The information is assumed to be given at each time, which means that the generated accessibility relation may change for different times. In this sense the proposed logical framework is both flexible and dynamic.

The information comes from the adaptation of different contexts by action of the second order or higher orders. The degree of significance of the worlds is proportional to the intensity of the adaptive field in the space of the worlds. We have defined the adaptive field in (8.6) as the field which compensates the asymmetry in the change of the space. The intensity of this field is the module of the tensor H. One example of this field in physics is the gravitational field the intensity of which compensates the change of the reference generated by the gravitational masses.

8.4   Definition of Accessibility Relations using Degrees of Significance and Adaptive Field

8.4.1 Fundamental Definitions

We can locate the possible worlds in a given reference space Γ. The positions of the worlds in Γ are arbitrary and we can create a set of worlds in this space. A sentence will be either true or false in each of the worlds created. Any world is associated with a function $g(\mathbf{x}; x_0(t))$, called *the Adaptive Field,* the maximum value of which is located at the $x_0 \in \Gamma$, the position of the world ω at a moment of time t. For simplicity we can use the Gaussian function for g. When Γ is the two dimensional space, the function g is:

$$g\,(x,y\,;x_0(t),\,y_0(t)) \;=\; \exp\{\,-\,\alpha(t)\,[\,(\,x - x_0(t)\,)^2 + (y - y_0(t)\,)^2\,]\,\} + K(t), \tag{8.7}$$

where $\alpha(t)$ and $K(t)$ are parameters at time t.   In this work we will call it an *elementary Adaptive Field.* When we know the time t, the parameter t will be omitted for simplicity.   Given the Adaptive Field SIG (ω_k) g $(\mathbf{x}\,;\,x_1(t)\,)$, for one world ω_k the total Adaptive Field for the set of worlds is given by the function:

$$F\,(\,\mathbf{x},t\,) = SIG\,(\omega_1)\,g\,(\,\mathbf{x}\,;\,x_1(t)\,) + ...+SIG(\omega_n)\,g\,(\mathbf{x}\,;\,x_n(t)\,) \tag{8.8}$$

Where SIG (ω_1), ..., SIG(ω_n) are the significance values located in the worlds $\omega_1,\ \ ,\omega_n$.

For any sentence p the Adaptive Field is:

$$F_p(\,\mathbf{x},t\,) = SIG\,(\,\omega_1,p\,)\,g\,(\,\mathbf{x}\,;\,x_1(t)\,) + ...+SIG(\,\omega_n\,,p\,)\,g\,(\mathbf{x}\,;\,x_n(t)\,) \tag{8.9}$$

Where SIG (ω_1, p), ..., SIG(ω_n, p) are the significance values located in the worlds ω_1, , ω_n for the proposition p. SIG (ω_1, p) = SIG (ω_k) when p is true in ω_k and SIG (ω_1, p) = 0 when p is false in ω_k.

For all the propositions $p_1, p_2,$., p_n the Adaptive Field is:

$$F (\mathbf{x}, t) = F_{p1} (\mathbf{x}, t) + ... + F_{pn} (\mathbf{x}, t) \qquad (8.10)$$

the total Adaptive Field is the superposition of the Adaptive Fields associated with any sentence or proposition p.

Definition: The Adaptive Field is formally defined as the following tuple:

$$S F = < \Gamma, W_t, F, \eta, \delta, S >, \qquad (8.11)$$

where Γ is the reference space, W_t is the set of worlds at time t , δ : $W_t \rightarrow \Gamma$ where δ is the function that at the time t associates one world with a point in the space Γ , F is the value of the Adaptive Field, η: $W_t \rightarrow F$ and S is the set of sources of the Adaptive Field. The sources S are the set of points where the field is generated.

In the works of Lewis [8], Dubois and Prade [3], Resconi, Klir, and St.Clair [15], the same structure is found where sentences are connected by real numbers. Any world of modal logic is associated with a couple consisting of a world and a real number in this way:

$\omega \rightarrow (\omega, a)$, where $a \in \mathbf{R}$ (where $\mathbf{R}$ is the set of real numbers)

For Lewis [8] the number "a" is a probability measure, for Dubois and Prade [3] the number "a" is a possibility measure, and Resconi, Klir, and St.Clair [15] consider the number "a" to be a generic weight assigned to any world to obtain different uncertainty measures. We stress that the Adaptive Field is an extension of the previous approaches to modal logic.

8.4.2 Accessibility Relations by the Adaptive Field

To obtain the Accessibility Relation in the Adaptive Field we use the continuity property of the Adaptive Field $F (\mathbf{x}, t)$. In fact by using the partial derivative $\dfrac{\partial F}{\partial x_i}$, we know the direction where the Adaptive Field decreases. For $\dfrac{\partial F}{\partial x_i} < 0$ the significance of the Adaptive Field decreases.

Definition:

For any two worlds there exists an accessibility relation from one world with a high value to the world of lesser significance.

From the previous definition we conclude that the accessibility relation in modal logic is associated with the sign of the partial derivative $\dfrac{\partial F}{\partial x_i}$ of the Adaptive Field.

Example:

Let the context C_t at the time t be given by Table 8.1.

Table 8.1. Space of the Worlds , Propositions, and SIG or degree of Significance which is Proportional to the Adaptive Field.

worlds	p_1 proposition one	p_2 proposition two	SIG significance
ω_1	1	0	1
ω_2	1	1	2
ω_3	0	1	1

In the Table 8.1 the space of the world $W = (\omega_1 , \omega_2 , \omega_3)$, the proposition p_1 is true in ω_1 and ω_2.

The proposition p_2 is true in ω_2 and ω_3 and the degree of significance is the sum of the true values in the worlds. With the two dimensional space $\Gamma = (X , Y)$ we locate the worlds as follows :

$$\delta (1,0) = \omega_1, \ \delta (0,1) = \omega_2, \ \delta (1,1) = \omega_3$$

The three elementary Adaptive Fields are:

$$g(x, y ; 1, 0) = \exp\{-\alpha [(x - 1)^2 + (y - 0)^2] \} + K, \text{ for the world } \omega_1$$

$$g(x, y ; 1, 1) = \exp\{-\alpha [(x - 1)^2 + (y - 1)^2] \} + K, \text{ for the world } \omega_2$$

$$g(x, y ; 0, 1) = \exp\{-\alpha [(x - 0)^2 + (y - 1)^2] \} + K, \text{ for the world } \omega_3$$

We obtain:

$$F_{p1} (x) = g(x, y ; 1, 0) + g(x, y ; 1, 1) , \ F_{p2} (x) = g(x, y ; 1, 1) + g(x, y ; 0, 1)$$

and,

$$F (x) = F_{p1} (x) + F_{p2} (x) = g(x, y ; 1, 0) + 2 g(x, y ; 1, 1) + g(x, y ; 0, 1)$$

For the accessibility relation we must compute the partial derivative:

$$\frac{\partial F}{\partial x} = -2\alpha[(x-1)\, g(x, y;\, 1,\, 0) + 2\,(x-1)\, g(x, y;\, 1,\, 1) + x\, g(x, y;\, 0,\, 1)]$$

$$\frac{\partial F}{\partial y} = -2\alpha[\, y\, g(x, y;\, 1,\, 0) + 2\,(y-1)\, g(x, y;\, 1,\, 1) + (y-1)\, g(x, y;\, 0,\, 1)]$$

when $\alpha = \frac{1}{2}$ we have the graphic representation of $F(\mathbf{x})$ shown in Figure 8.2.

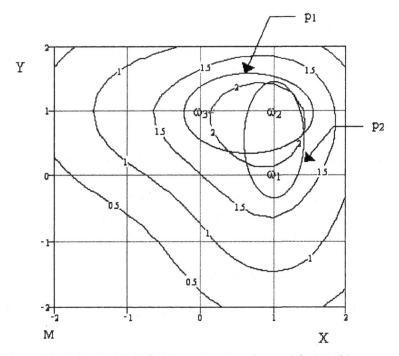

Figure 8.2. Adaptive Field for Sentences p_1 and p_2 and for Worlds ω_1, ω_2, and ω_3.

Figure 8.3 shows the vector field of partial derivatives for the three worlds ω_1, ω_2, and ω_3 in the two-dimensional space Γ.

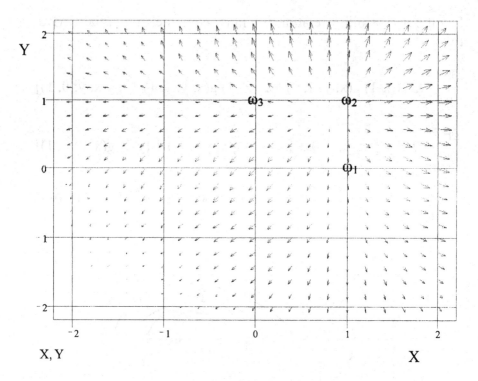

Figure 8.3. Vector Field of Partial Derivatives of the Adaptive Field.

Given the sentence p with the set $\|\,p\,\| \in P(W)$, any set $\|\,p\,\|$ can be associated with an Adaptive Field by the formula (8.10) . When we know the form of the field $F(\mathbf{x})$ we can give a modal logic image of $F(\mathbf{x})$ (adaptive field) as follows. Given a field $F(\mathbf{x})$ we can expand it in the elementary field $g\,(\mathbf{x}\,;\,\mathbf{x}_k\,)$:

$$F(\mathbf{x}) = c_1\, g\,(\mathbf{x}\,;\,\mathbf{x}_1\,) + \ldots\ldots\ldots + c_n\, g\,(\mathbf{x}\,;\,\mathbf{x}_n\,) \tag{8.12}$$

If,

(1) at any position x_i we put a world and

(2) when $c_1,.....,c_n$ are degrees of significance, we can then represent the field F(x) as an Adaptive Field.

From the worlds and propositions given in Table 8.1 we can generate an Adaptive Field. From this field we can generate the significance in the set of the worlds.

8.4.3 Set Theory and the Adaptive Field

Given two classes $A = \{G_1, ..., G_n\}$ and $B = \{B_1,..., B_m\}$ of subsets of X. Then B is a *base* for A if and only if:

(1) $B_1 \cup .... \cup B_m = X$; and for every couple consisting of sets B_i, B_j $\in B$, $B_i \cap B_j = \varnothing$ and

(2) every set $G_i \in A$ is the union of the members of B.

From the two classes A and B we can define the set of worlds W and the set of sentences $ATOMS_t$, where $ATOMS_t$ is the set of sentences at the time t, in the following *global way*: every member B_i of B is regarded as a world ω_i; and also at every element G_j of A we assume that there is a sentence p_j such that $\| p_j \| = Gj$. From A and B we can also define the function:

$f : W \times ATOMS_t \rightarrow \{1, 0\}$ by
$f(\omega_i, p_j) = 1$ if $B_i \subseteq G_j$ with $B_i \in B$ and $G_j \in A$,
$f(\omega_i, p_j) = 0$, otherwise.

Example:

The class of sets $A = \{\{a,b\}, \{a,b,c\}, \{a,b,c,d\}\}$ has the base:

$B = \{\{a,b\}, \{c\}, \{d\}\}$

We then have the following global worlds, sentences and valuation function described in Table 8.2.

Table 8.2. Global Worlds, Sentences and Valuation Function obtained by Set Theory.

Global worlds	$p_1 \equiv \{a,b\}$	$p_2 \equiv \{a,b,c\}$	$p_3 \equiv \{a,b,c,d\}$	SIG
$\omega_1 \equiv \{a,b\}$	1	1	1	3
$\omega_2 \equiv \{c\}$	0	1	1	2
$\omega_3 \equiv \{d\}$	0	0	1	1

We locate the worlds in a two dimensional space Γ as follows:

$$\delta(1,0) = \omega_1 , \delta(0,1) = \omega_2, \delta(1,1) = \omega_3$$

the three elementary Adaptive Fields at time t are:

$g(x, y ;1,0) = \exp\{-\alpha[(x-1)^2 + (y-0)^2]\} + K$ for the world $\omega_1 \equiv \{a,b\}$

$g(x, y ;1,1) = \exp\{-\alpha[(x-1)^2 + (y-1)^2]\} + K$ for the world $\omega_2 \equiv \{c\}$

$g(x, y ; 0,1) = \exp\{-\alpha[(x-0)^2 + (y-1)^2]\} + K$ for the world $\omega_3 \equiv \{d\}$

We obtain:

$F_{p1}(x) = g(x, y ;1,0)$
$F_{p2}(x) = g(x, y ; 1,0) + g(x, y ;1,1)$
$F_{p3}(x) = g(x, y ; 1,0) + g(x, y ;1,1) + g(x, y ; 1,0)$

And,

$F(x) = F_{p1}(x) + F_{p2}(x) + F_{p3}(x) = 3\, g(x, y ;1,0) + 2\, g(x, y ;1,1) + 1\, g(x, y ;0,1)$

8.4.4 Laws, Facts and the Adaptive Field

Sowa [20] pointed out that philosophers since Aristotle have recognized that modality is related to laws. We recognise that the modal logic is based on the modality expressed by the definition of possible world. According to Sowa [20], Dunn introduced the possible world as a pair of facts and laws defined by <F, L>. Here F are the sentences included in the world called *facts* of a possible world and L is a subset of F called the *laws* of that particular world.

Sowa gives an effective method to introduce the possible world. Facts are the possible behaviour or actions of human beings. Laws are the behaviours or actions that are accepted in a specific society. The possible world contains the rules L and facts that one society uses for its necessity. When we change the possible facts and possible laws, we change the environment where one society lives and also the limitation imposed by the society laws. Sowa describes a society with a very strong logic model and at the same time gives a concrete example of the power of the modal logic itself.

Any world $< F , L >$ can be denoted by ω^D, at any two pairs $< F_1 , L_1 >, < F_2 , L_2 >$ Dunn defined an accessibility relation from the world $\omega^D{}_1$ and $\omega^D{}_2$ where the laws L_1 are a subset of the facts F_2. We know that every law is a subset of possible facts. The law establishes a limitation on the possible facts. When only facts F_2 of the second worlds are included in the laws L_1, we say that there exists an accessible relation between the two worlds. The constrain in the worlds L_1 selects facts that are included in the facts of the second world. The second world facts are an extension of the possible facts defined by the laws L_1. In a formal way:

$$R\,(\omega^D{}_1 , \omega^D{}_2) \equiv L_1 \subset F_2 \tag{8.13}$$

We now compare our definition of the accessibility relation with the Dunn's definition in (8.11). Given a class of sets $A = \{F_1, F_2,$

L_1, L_2 } with $L_1 \subset F_2$, let us consider the following base class B of sets:

$$B = \{ L_1 \cap L_2^c, L_1 \cap L_2, L_2 \cap L_1^c \cap F_1, L_2 \cap L_1^c \cap F_2 \} \}$$

by which the set of worlds is associated with the base in this way:

$$\omega_1 \equiv L_1 \cap L_2^c, \ \omega_2 \equiv L_1 \cap L_2, \ \omega_3 \equiv L_2 \cap L_1^c \cap F_1, \ \omega_4 \equiv L_2 \cap L_1^c \cap F_2$$

Where L^c is the set of facts that are not allowed for the law L. These facts are the complementary facts possible in a specific society. So we have the context by laws and facts in Table 8.3.

Table 8.3. Laws and Facts in a Table of the Worlds.

worlds	F_1	F_2	L_1	L_2	SIG
ω_1	1	1	1	0	3
ω_2	1	1	1	1	4
ω_3	1	1	0	1	3
ω_4	0	1	0	1	2

In Table 8.3 the set of facts are $F_1 = \{\omega_1, \omega_2, \omega_3 \}$ and $F_2 = \{\omega_1, \omega_2, \omega_3, \omega_3\}$ the set of laws are $L_1 = \{\omega_1, \omega_2\}$ and $L_2 = \{\omega_2, \omega_3, \omega_3\}$ where the worlds $\{\omega_1, \omega_2, \omega_3, \omega_3\}$ are different from the worlds defined by Sowa or ω^D The worlds ω_k are the elementary facts.

The worlds ω_k and the set of Facts and laws can also be shown as in Figure 8.4.

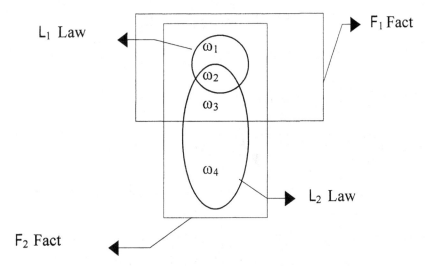

Figure 8.4. Worlds in Laws and Facts.

The worlds ω^D_1 and ω^D_2 using the definitions given by Dunn are :

$$\omega^D_1 \equiv <F_1 , L_1> \equiv \{ \omega_1 , \omega_2 \} \text{ and}$$
$$\omega^D_2 \equiv <F_2 , L_2> \equiv \{ \omega_2 , \omega_3 , \omega_4 \}$$

For the definition 8.11 and because L_1 is included in F_2, $R (\omega^D_1 , \omega^D_2)$ is true, ω^D_2 is accessible from ω^D_1. Because L_2 is not included in F_2, ω^D_1 is not accessible from ω^D_2. Using Table 8.3, and the accessible relation obtained by the adaptive field we have the accessible diagram in Figure 8.5.

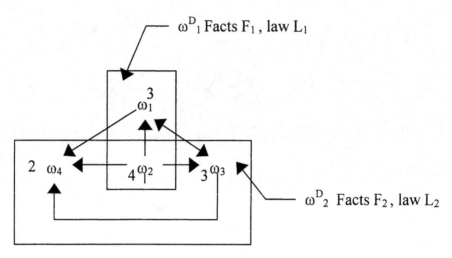

Figure 8.5. Accessibility Relation using Adaptive Field.

In Figure 8.5 the arrows which move from ω^D_1 to ω^D_2 are four, the arrows which move from ω^D_2 to ω^D_1 are two. Because the number of arrows which go from ω^D_1 to ω^D_2 is greater than the number of arrows which go from ω^D_2 to ω^D_1, we can say that ω^D_2 is accessible from ω^D_1 but that the opposite is not true. With this simple example we show that there exists a bridge between the definition of accessibility given in 8.11 and the definition of accessibility obtained by the adaptive field.

8.4.5 Scott-Montague Models and Adaptive Field

Let $A = \{G_1,....., G_n\}$ be a class of subsets of a given set X. At any G_j exists a sentence p_j such that:

$$\| p_j \| = G_j \tag{8.14}$$

We can write $A = \{\|p_1\|,....., \|p_n\|\}$. We can define from A a class B of subsets of X that form a *base* for the class A. In Montague Models we define the set of worlds at the time t as the class B and so:

$W_t = B$ $\qquad\qquad\qquad\qquad\qquad\qquad\qquad\qquad$ (8.15)

We can obtain the function N_A by:

$N_A(\omega) \equiv \{ G_i \mid \omega \subseteq G_i \}$ $\qquad\qquad\qquad\qquad\qquad$ (8.16)

Given the world ω, the function $N_A(\omega)$ gives all the elements of A which are $\{G_1,....., G_n)$ that include the given world. We can then formulate a Scott-Montague model $< W_t, N_A, V_t >$ as the result of the previous definition of the set of worlds, the function N_A and the evaluation V_t by $\| p_j \| = G_j$ where $\| p_j \|$ is the set of worlds in which p is true.

With the function $N_A(\omega)$ we can define the accessible relation between two worlds as follows:

$\omega_k \ R \ \omega_h$ when $\omega_k \in \cap N_A (\omega_h)$ $\qquad\qquad\qquad$ (8.17)

Example:

Given a class of sets $A = \{\{a\}, \{a,b \}, \{a,b,c \},\{a,b,c,d \}\}$ where $G_1 = \{a\}$, $G_2 = \{a,b \}$, $G_3 = \{a,b,c \}$, $G_4 = \{a,b,c,d \}\}$ we can generate its base by the set of intersections:

$G_1 \cap G_2 \cap G_3 \cap G_4 = \{a\}$ $\quad$ $G_1^c \cap G_2 \cap G_3 \cap G_4 = \{b\}$

$G_1 \cap G_2^c \cap G_3 \cap G_4 = \varnothing$ $\quad$ $G_1^c \cap G_2^c \cap G_3 \cap G_4 = \{c\}$ and so on. In conclusion all the intersections give the class of sets $\{\{a \}, \{b \}, \{c \}, \{d \}\}$ that is the base B searched. For the definition of worlds in 8.15 we obtain:

$W = B = \{\{a \}, \{b \}, \{c \}, \{d \}\}$

For the (8.13) the proposition p_1 is true in G_1 so it is true only in one world $\omega_1 = \{a \}$; p_2 is true in G_2 that is the composition of the sets $\{a\} \cup \{b\} = G_2$, so p_2 is true in the worlds $\omega_1 = \{a\}$ and $\omega_2 = \{b\}$; p_3 is true in G_3 that is the composition of the sets $\{a\} \cup \{b\} \cup \{c\} = G_3$ so p_3 is true in the worlds $\omega_1 = \{a\}$ and $\omega_2 = \{b\}, \omega_3 = \{c\}$; p_4 is true in G_4 that is the composition of the sets $\{a\} \cup \{b\} \cup \{c\} \cup \{d\} = G_4$ so p_4 is true in the worlds $\omega_1 = \{a\}$ and $\omega_2 = \{b\}, \omega_3 = \{c\}$ and $\omega_4 = \{d\}$. When we replace True with the value 1 and False with the value 0, we collect all the previous information in the Table 8.4 where we compute also the significant value to build the adaptive field.

Table 8.4. Context to generate a Scott-Montague model.

	p_1	p_2	p_3	p_4	SIG
ω_1	1	1	1	1	4
ω_2	0	1	1	1	3
ω_3	0	0	1	1	2
ω_4	0	0	0	1	1

For the (8.15) given the world $\omega_1 = \{a\}$ we remark that:

$\{a\} \subseteq \{a\} = G_1 \equiv p_1$,
$\{a\} \subset \{a,b\} = G_2 \equiv p_2$,
$\{a\} \subset \{a,b,c\} = G_3 \equiv p_3$,
$\{a\} \subset \{a,b,c,d\} = G_4 \equiv p_4$,

So $N_A(\omega_1) = \{p_1, p_2, p_3, p_4\}$. In the same way we obtain the values for the other worlds and we obtain:

$N_A(\omega_1) = \{p_1, p_2, p_3, p_4\} = \{G_1, G_2, G_3, G_4\}$,
$N_A(\omega_2) = \{p_2, p_3, p_4\} = \{G_2, G_3, G_4\}$
$N_A(\omega_3) = \{p_3, p_4\} = \{G_3, G_4\}$ $N_A(\omega_4) = \{p_4\} = \{G_4\}$

We can then formulate a Scott-Montague model $< W_t, N_A, V_t >$. For the definition of the accessible relation in (8.17) we have:

$\cap N_A(\omega_1) = G_1 \cap G_2 \cap G_3 \cap G_4 = \{a\}$
$\cap N_A(\omega_2) = G_2 \cap G_3 \cap G_4 = \{a, b\}$
$\cap N_A(\omega_3) = G_3 \cap G_4 = \{a, b, c\}$
$\cap N_A(\omega_4) = G_4 = \{a, b, c, d\}$

So we have the accessibility diagram in Figure 8.6 where all the worlds are accessible to themselves.

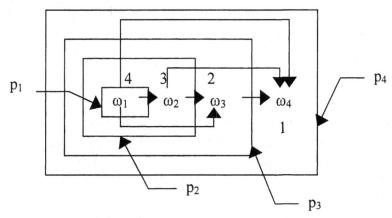

Figure 8.6. Accessibility relation generated by the function N_A equal to the accessibility relation by the adaptive field.

8.5 Uncertainty by Adaptive Field

8.5.1 Hierarchical uncertainty Metatheory

In a series of papers [4, 6, 15, 16, 17, 18], Resconi et al. have developed a new approach to uncertainty. The new approach is based on Kripke models for modal logic. We know that the Kripke model is given by the structure:

$$< W, R, V > \tag{8.18}$$

It is suggested in the new approach that a real number should be added to the former model in such a way as to obtain the model:

$$< W, R, V, \Psi > \tag{8.19}$$

where Ψ is included in the set of real numbers. Any world is associated with one real number. With the model given in (8.19) we can build a *hierarchical uncertainty metatheory* in which it is possible to calculate an expression for the membership function in fuzzy set theory as follows:

$$\mu_X(x) = \sum_k \eta(\omega_k)V(p_x, \omega_k), \tag{8.20}$$

where $\eta_k(\omega_k) \in \Psi$, $\omega_k \in W$, $p_x =$ " x belongs to the set X " and $V(p_k, \omega_k)$ is the valuation function in the Kripke model. With this model (8.19) we can also obtain the expression for the belief and plausibility measures in the Dempster –Shafer evidence theory using the expressions:

$$Bel(D) = \sum_k \eta(\omega_k)V(\square\, p_D, \omega_k) \quad \text{and}$$

$$Pl(D) = \sum_k \eta(\omega_k)V(\lozenge\, p_D, \omega_k), \tag{8.21}$$

where p_A is " A given incompletely characterized element ε which is classified in set D". Using the previous uncertainties we can obtain other uncertainties with a repetition of the Necessity and the Possibility operators. It is also possible to use different types of necessity and possibility operators (Chellas [1]).

We will show in this section that the Hierarchical Uncertainty Metatheory based upon modal logic using an Adaptive Field built by using Degrees of Significance is the basic mathematical structure which can connect Logic with Uncertainty. Considering the information expressed by set theory, we can obtain uncertainty from information given by the Hierarchical uncertainty Metatheory and the Adaptive Field.

8.5.2 Conditional Probability Measure and the Adaptive Field

Dubois and Prade [3] have recently suggested a new semantics for the conditional object q|p defined in Lehmann's system. Within the new semantics a conditional object q|p is considered as a family of possibility measures Π agreeing with the constraint:

$$\Pi\,(\,p \wedge q\,) > \Pi\,(\,p \wedge \neg\, q\,). \tag{8.22}$$

In the representation of the probability by the model $M = < W_t\,, I_t\,, \Psi >$, we define degrees of significance that are connected with the Adaptive Field. It is noted that the knowledge encoded by using the degree of significance and the Adaptive Field is different from the knowledge encoded by using probability. Using the degree of significance and the Adaptive Field we described, in the normal course of events, which worlds are more likely and plausible than others. This can also be connected with possibility theory [2] and normality ranking. In possibility theory every event or nested sets are valuated with a possible value which gives the degree of possibility for the event. We have a criterion to decide if one event is more possible than another. With an example we can show, using the significance (adaptive field), the same inequality (8.20) that defines the conditional probability q|p.

Example:

When p is true in the worlds $\{\omega_1, \omega_2 \}$ and q is true in the worlds $\{\omega_2, \omega_3 \}$ and we substitute True with 1 and False with 0, we can compute the significance equal to the possibility measure Π as shown in Table 8.5.

Table 8.5. Worlds and Propositions.

worlds	$p \equiv \{\omega_1, \omega_2\}$	$q \equiv \{\omega_2, \omega_3\}$	SIG	Π
ω_1	1	0	1	1
ω_2	1	1	2	2
ω_3	0	1	0	0

Using Table 8.5 we obtain:

$$\Pi(\omega_2) > \Pi(\omega_1) \tag{8.23}$$

where $\omega_2 \equiv p \wedge q$, $\omega_1 \equiv p \wedge \neg q$. Using (8.21) we can find the same inequality in Dubois and Prade [3]. In this way the conditional probability can be developed to obtain a clearer position in the uncertainty domain as shown in Dubois Prade Possibility Theory by using inequality (8.20) and by the significance with inequality (8.21) that is another representation of the same inequality (8.20).

8.6 Adaptive Agents and Adaptive Field

When in one context we are unable to compensate for every rule that comes from other contexts, we have studied in chapter 6 a fuzzy logic description of incoherence. In this chapter we study the possibility that for every channel or world which transports a set of rules to the context there is an associated degree of significance or adaptive agent. In Figure 8.7 we show five worlds one for every context each with its own set of rules and degree of significance. Among the worlds there is an accessible relation that connects the worlds having a high significance with worlds of lower significance. Using accessibility relationships we can compute the belief measure and plausibility measure. These both are fuzzy measures. Refer to the expression (8.19). In this chapter and chapter 6 we are able to study fuzzy sets and fuzzy measures to determine the incoherence within a context with different rules.

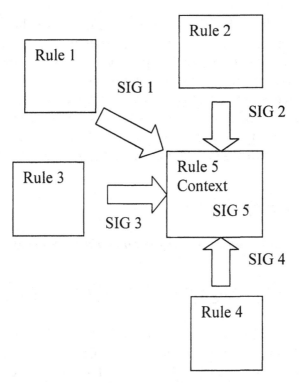

Figure 8.7. The degree of significance SIG or adaptive field for different contexts.

8.7 Conclusion

In this chapter we have introduced the Adaptive Field in a general way. The Adaptive Field is not a physical field but is a virtual and logic field that compensates for asymmetry generated by variations in the reference within the space where the worlds are located.

It is done in such a way as to change the contexts of worlds to obtain symmetry or coherence. The adaptive field acts inside a conceptual space of the worlds and we can decide if a sentence is possibly or necessarily true or false (accessible relation). The intensity of the adaptive field is proportional to the significance of the world.

In the worlds where the Adaptive Field is of zero value the incoherence or asymmetry for the change of reference is equal to zero. In worlds where the effort to compensate for asymmetry is maximum, the worlds are of maximum significance. Every world w with the significance S can access all worlds w* where the adaptive field has a significance of H ≤ S. That is when a sentence is false in w but is true in w* the sentence is possibly true. In conclusion the adaptive field establishes a relationship among different worlds.

Bibliography

1. Chellas B.F., *Modal Logic: An Introduction*. Cambridge University Press, 1980.

2. Dubois D. and Prade H., *Possibility Theory*. Plemun Press, 1988.

3. Dubois D. and Prade H., Conditional objects, possibility theory and default rules. In G.Crocco, L.Farinas del Cerro, and A.Herzig, Conditionals: from Philosophy to Computer Science, Studies in Logic and Computation 5, pp.301-336, 1995.

4. Harmanec D., Klir G.J. and Resconi G., "On Modal Logic Interpretation of Dempster-Shafer Theory of Evidence". *Int. J. Intelligent Systems*, vol.9, pp.941-951, 1994.

5. Hughes G.E. and Cresswell M.J., *An Introduction to Modal Logic*. Methuen, 1968.

6. Klir G.J. and Harmanec D., "On Modal Logic Interpretation of Possibility Theory". *Int. J. of Uncertainty, Fuzziness and Knowledge-Based Systems*, vol.2, pp.237-245, 1994.

7. Klir G.J. and Yuan B., *Fuzzy Sets and Fuzzy logic*. Prentice Hall, 1995.

8. Lewis D., "Probabilities of Conditionals and conditional proba-bilities". Philosohpical Review, vol.85, pp.297-315, 1976.

9. Mignani R., Pessa E. and Resconi G., "Non-Conservative Gravi-tational Equations, General Relativity and Gravitational", vol.29, no.8, pp.1049-1073, 1997.

10. Muirhead H., *The Physics of Elementary Particles*. Pergamon press, 1968.

11. Murai T., Nakata M. and Shimbo M., Ambiguity, Inconsis-tency, and possible-worlds: a new logical approach. Proceeding of the Ninth Conference on Intelligence Technologies in Hu-man-Related Sciences, Leòn, Spain, pp.177-184, 1996.

12. Murai T., Kanemitsu H. and Shimbo M., "Fuzzy sets and bi-nary-proximity-based rough sets", *Information Science*, vol.104, pp.49-80, 1998.

13. Pauli W., Relativitatstheorie. In Encyklopadie der mathematishen Wissenschaften, vol. 5, Article 19, Teubner Lipsia, 1921.

14. Resconi G. and Jessel M., "General System Logical Theory", *Int. J. General Systems*, vol.12, pp.159-182, 1986.

15. Resconi G., Klir G.J. and St.Clair U., "Hierarchical Uncertainty Metatheory Based upon Modal Logic". *Int. J. of General Sys-tems*, vol.21, pp.23-50, 1992.

16. Resconi G., Klir G.J., St. Clair U. and Harmanec D., "On the Integration of Uncertainty Theories". *Int. J. of Uncertainty, Fuzziness, and Knowledge-Based Systems*, vol.1, pp.1-18, 1993.

17. Resconi G. and Rovetta R., "Fuzzy Sets and Evidence Theory in a Metatheory Based upon Modal Logic". Quaderni del Seminario Matematico di Brescia n.5 ,1994.

18. Resconi G., Klir G.J., Harmanec D. and St. Clair U., "Interpretations of Various Uncertainty Theories Using Models of Modal Logics: A Summary". Fuzzy Sets and Systems, vol.80, pp.7-14, 1996.

19. Shafer G., *A Mathematical Theory of Evidence*. Princeton University Press, 1976.

20. Sowa J.F., *Knowledge Representation: Logical, Philosophical, and Computational Foundation*. PWS Publishing Co. Pacific Grove, CA. 1999.

21. Wille R., Lattices in data analysis: how to draw them with a computer. In I.Rival(ed.): *Algorithms and order,* Kluwer, Dordrecht-Boston, pp.33-58, 1989.

22. Zadeh L.A., "Fuzzy Sets. Information and Control", vol.8, pp.338-353, 1965.

23. Zadeh L.A., "A Simple View on the Dempster-Shafer Theory of Evidence and its Implication for the rule of Combination". *AI Magazine*, vol.7, pp.85-90, 1986.

Chapter Nine

Adaptive Agents in the Physical Domain

This chapter presents the study of Physical Phenomena in line with the adaptive agent described in the previous chapters. We have demonstrated that in the model of the physical reality we can find the same logical structure as that used for more conceptual and abstract fields of research.

9.1 Virtual Sources as Compensation and Real Sources as Adaptation

A serious drawback is hidden behind the apparent perfection of classical mechanics: that is the need of introducing external sources, as an indispensable tool for forecasting the details of dynamical evolution of mechanical systems and other cognitive structures. Concerning traditional classical mechanics, where the only macroscopic field is substantially the gravitational field, without the introduction of sources we would obtain trivial models. These would be unable to account for experimental data, such as for example relative to planetary motion.

The introduction of suitable external sources is only a process to adapt models to fit the observed data. It is a measure of our inability to forecast all possible phenomena from a global point of view. Lacking an explanation of an observed dynamical trajectory, we add an external source in a suitable space-time region to reconcile the structure of our model with the observations. For this reason, the introduction of external sources into a classical mechanics model cannot be considered as a true process of "compensation" for our ignorance, or lack of information for the system under consideration. This is becasue it is not part of a well-defined transformation between our old model and the new model. On the contrary, the introduction of these external sources is merely an "adaptation" leading to a model logically different from the one without added external sources. These added data are essentially local, and do not take account of the true nature of the model and its inherent transformations. The sources can generate the field but the reverse process is not physically valid. These added sources are different from the essential field and give the possibility that the field may be changed by the data (sources) from the other physical or logical domains. For example, when the field of sound is generated by a loud-speaker which is controlled by the electromagnetic device, the source is the bridge between two completely different domains of physics, electricity and sound waves.

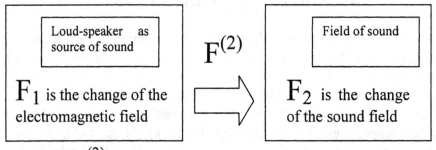

$F^{(2)}$ is the action of the second order that transforms the change of the electromagnetic field into the change of the sound field

Figure 9.1. Context of the Sources of Sound with its Rules and Context of the Field of Sound. The two contexts are connected by a function at the second order or second order adaptive action.

In the first instance the introduction of sources by hand gave rise to good performance of models – similar to those in classical mechanics when forecasting experimental data. It is well known that the successful results are obtained by the use of classical mechanics when applied to describe the motions of celestial objects, of solid bodies and of classical fluids. Notwithstanding these results, the drawbacks of this procedure appeared at the beginning of last century mainly in connection with the discovery of electric and magnetic phenomena. These drawbacks are of two types:

(1) The introduction of external sources gives rise to models containing two entities of different logical natures: the fields and the sources, the evolution of which is based on well-defined rules. These constitute the essence of models and sources. The evolution and nature of these models is absolutely unpredictable within the models utilised. This circumstance precludes any solution to the problem of the definition of what the sources are. For example, what is the logical difference between a mass and an electric charge?

(2) A theory containing external sources works only if we neglect any interaction between these sources. Any self-interaction between the source and the field created by the source itself (non linear phenomena) must be neglected. This requirement results in a theory which cannot treat phenomena such as phase transitions or reaction radiation.

These difficulties led to the birth of a new approach to field theory. This emphasizes the role of fields as primary entities and attempt to reduce external sources only to the regions characterized by high field strengths. Scientists such as Faraday pioneered this approach. The validity was, in a certain sense, confirmed by the discovery of electromagnetic waves. Here the displacement current produced by the field itself acts as if it were an external source of magnetic field.

9.2   Compensation Current or Internal Adaptive Current in the Electromagnetic Field

Given the magnetic field H, we know that for the circulation theorem we have that the current J (source) is equal to:

$$J = \text{rot } H$$

Where rot is the differential operator "rotor". From the previous relation we have an internal rule at the source for which:

$$\text{div } J = 0$$

where div is the differential operator divergence. The previous rule comes from the property:

$$\text{div rot } H = 0$$

So we have the $F^{(2)}$ or ELS(2) (second order action described in chapters 2 and 3).

$$F^{(2)} : F_1 \rightarrow F_2 \text{ where } F_1 : J_1 \rightarrow J_2 \text{ and } F_2 : \text{div } J_1 \rightarrow \text{div } J_2.$$

Where J_1 is the current for low frequency for which div $J_1 = 0$, and J_2 is the current for high frequency where:

$$\text{div} J_2 + \frac{\partial \rho}{\partial t} = 0$$

is the conservation equation for the charge. For the Gauss theorem we have,

$$\text{div } \varepsilon\, E = \rho \quad (\text{where E is the intensity of the electric field.})$$

and,

$$div(J_2 + \varepsilon \frac{\partial E}{\partial t}) = 0.$$

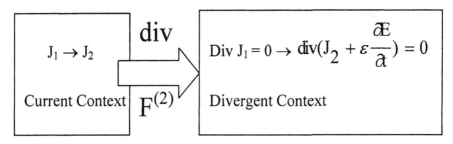

Figure 9.2. Context of the Currents and Context of the Divergence of the Current. In the Divergent Context we must add, for the physical symmetry, the compensation term $\varepsilon \frac{\partial E}{\partial t}$.

From Maxwell the displacement current $J = \varepsilon \frac{\partial E}{\partial t}$, which is dependent on the frequency of the current in a wire. When the frequency is low the displacement current is low and when the frequency is high the displacement current is high and electromagnetic waves are generated. The current J is not produced by physical particles but is generated by the electric field and the change in the magnetic field. The current J joins the electric field with the magnetic field for large changes in the electric field. The current $\varepsilon \frac{\partial E}{\partial t}$ is not an "adaptive" source generated by an external domain, but is a "compensation" or internal adaptation within the domain that eliminates the conflict between the static electromagnetic phenomena and the phenomena due to field changes in time.

9.3 Compensation Sources or Internal Adaptive Sources for Sound

Another use of the sources as "compensation or internal adaptation" was given by M. Jessel. When we want to reshape a field for a special task, we "compensate" the original source by using secondary sources that are dependent on the field itself.

Given that E is the field and λE is the reshaped field. The formula relating the sources and the field is as follows:

$$(\frac{\partial^2}{\partial x^2} - \frac{1}{c_1^2} \frac{\partial^2}{\partial t^2}) E = S \tag{9.1}$$

Where c is the wave velocity in the field E and S is the wave. The equation for the reshaped field is:

$$(\frac{\partial^2}{\partial x^2} - \frac{1}{c^2} \frac{\partial^2}{\partial t^2}) \lambda E = \lambda S + S_\lambda$$

where,

$$S_\lambda = (\frac{\partial^2}{\partial x^2} - \frac{1}{c^2} \frac{\partial^2}{\partial t^2}) \lambda E - \lambda (\frac{\partial^2}{\partial x^2} - \frac{1}{c^2} \frac{\partial^2}{\partial t^2}) E = (\frac{\partial^2 \lambda}{\partial x^2} - \frac{1}{c^2} \frac{\partial^2 \lambda}{\partial t^2}) E$$

So in the ELS (2) (see Chapter 3) or second order action OP = $(\frac{\partial^2}{\partial x^2} - \frac{1}{c_1^2} \frac{\partial^2}{\partial t^2})$ is the operator that joins the context of the field E with the context of the sources S. For a desired new field λE it is not sufficient to simply reshape the source S in λS but we must include additional terms to avoid conflict. It is necessary to add a compensation term S_λ. S_λ will depend on E and on the reshaping function $\lambda(x,t)$. The source S_λ is therefore not a simple adaptation term without any meaning, but is an internal adaptation computed to obtain the coherent condition.

We know that in the second order action:

$$F^{(2)} : F_1 \rightarrow F_2$$

We can see in Figure 9.3 the elements:

$$\mathbf{F_1} : E \rightarrow \lambda(x,t)\, E \text{ and } \mathbf{F_2} : \left(\frac{\partial^2}{\partial x^2} - \frac{1}{c_1^2}\frac{\partial^2}{\partial t^2}\right) E \rightarrow \left(\frac{\partial^2}{\partial x^2} - \frac{1}{c_1^2}\frac{\partial^2}{\partial t^2}\right)$$

$$\lambda(x,t)\, E., \text{ OP} = \left(\frac{\partial^2}{\partial x^2} - \frac{1}{c_1^2}\frac{\partial^2}{\partial t^2}\right)$$

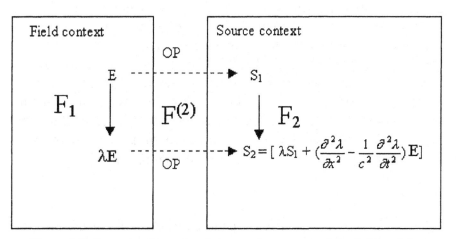

Figure 9.3. Second order action between the context of the field and the context of the sources where $\left(\dfrac{\partial^2 \lambda}{\partial x^2} - \dfrac{1}{c^2}\dfrac{\partial^2 \lambda}{\partial t^2}\right) E$ is the compensation term to obtain the coherence in the second order action $F^{(2)}$. The operator $\text{OP} = \left(\dfrac{\partial^2}{\partial x^2} - \dfrac{1}{c_1^2}\dfrac{\partial^2}{\partial t^2}\right)$ transforms the Field context into the Source context in (9.1).

9.4 Compensation or Internal Adaptation for the Electromagnetic Field

Another example of the "compensation" or "internal adaptation" is the compensation in the context of the electromagnetic field. The need for taking into account the electromagnetic field in the description of the motion of a charged particle showed that the structure of classical mechanics, even in absence of external sources, was totally unsuitable in its original form to represent a class of physical phenomena wider than that described in the eighteenth century.

However the formal structure of classical mechanics is appealing and very convenient from the mathematical point of view. In this situation there is a need for an operation able to maintain the formal structure of classical mechanics, and to add some new "compensating terms" to permit the study of new phenomena. In order to do such operation in the most efficient way, we need a precise definition of the new domain, so that the compensating terms can be computed according to well-defined rules when we want to maintain the formal apparatus of analytical mechanics used to study the motion of a charged particle in presence of an electromagnetic field. This requires that the ordinary momentum p_μ should be replaced by a new "generalized" momentum $p_\mu = m \dfrac{dx_\mu}{dt} + \dfrac{e}{c} A_\mu$. Here "e" is the particle electric charge and A_μ is the usual electromagnetic four-potential where $\dfrac{e}{c} A_\mu$ is the compensating term in the mechanical context. We will show the compensation effect also in the domain of mechanical energy.

$P_\mu = m \dfrac{dx_\mu}{dt}$ the total energy is:

$$W = \frac{1}{2m}\sum_{\mu}(\frac{dx_{\mu}}{dt})^2 + U = \frac{1}{2m}\sum_{\mu}(P_{\mu})^2 + U$$

Where m is the mass of the particle and U is the potential of the field where the particle moves.

With the electromagnetic field:

$$p_{\mu} - \frac{e}{c}A_{\mu} = m\frac{dx_{\mu}}{dt}$$

the "compensation" formula for the total energy is:

$$W = \frac{1}{2m}\sum_{\mu}(\frac{dx_{\mu}}{dt})^2 + U$$

$$= \frac{1}{2m}\sum_{\mu}(p_{\mu} - \frac{e}{c}A_{\mu})^2 + U$$

Using the language of the adaptive agent we obtain:

$$F^{(2)} : F_1 \rightarrow F_2$$

Where,

$$F_1 : p_{\mu} \rightarrow \frac{1}{2m}\sum_{\mu}(p_{\mu})^2 + U \text{ and}$$

$$F_2 : p_{\mu} - \frac{e}{c}A_{\mu} \rightarrow \frac{1}{2m}\sum_{\mu}(p_{\mu} - \frac{e}{c}A_{\mu})^2 + U$$

Where F_1 moves from momentum to energy in the context without electromagnetic field where we use the classical mechanics. F_2 is in the context with the electromagnetic field where we can restore the original symmetry in mechanics with the introduction of the compensation term A_μ. The original symmetry is given by the use of the traditional rules in classical mechanics.

9.5 Space –Time Transformation in a Physical Context

We will now take into consideration another form of compensation which is described as "local". It arises when we need a scientific theory to keep a given external form by using local transformations the parameters of which depend on space-time coordinates.

Using this compensation process we can give a meaning to the equation:

$$p_\mu - \frac{e}{c} A_\mu = m \frac{dx_\mu}{dt}$$

The derivative $\partial_\mu = \dfrac{\partial}{\partial x_\mu}$ (where μ=1, 2, 3, 4) is the index for the space – time coordinates. It is considered in all the contexts to be invariant. When this hypothesis changes, we create other contexts where the derivatives change. If the transformation T of the contexts is global, then all the points in the space- time change in the same way as the Galileo or Lorenz transformation. However the definition of the derivative cannot change. When T(x,t) is a function of space – time and when we use the same derivative in two contexts we obtain:

first context $\partial_\mu \psi = \dfrac{\partial \psi}{\partial x_\mu}$

second context $\partial_\mu T\psi = \dfrac{\partial T}{\partial x_\mu}\psi + T\dfrac{\partial \psi}{\partial x_\mu}$

When using the transformation T we transport the derivative from the first context to the second context we have two different derivatives in the same context:

$$T\partial_\mu\psi = T\frac{\partial \psi}{\partial x_\mu} \text{ and}$$

$$\partial_\mu T\psi = \frac{\partial T}{\partial x_\mu}\psi + T\frac{\partial \psi}{\partial x_\mu}$$

This is impossible because in the same context we must have only one derivative. Here the second order action is in a conflicting situation for the "local" transformation T(x,t).

To compensate for this conflict, we must change the definition of the derivative in the second context. The new derivative D_μ in the second context is:

$$D_\mu = \frac{\partial}{\partial x_\mu} - \frac{\partial T}{\partial x_\mu}$$

So

$$D_\mu T \psi = \frac{\partial T}{\partial x_\mu}\psi + T\frac{\partial \psi}{\partial x_\mu} - \frac{\partial T}{\partial x_\mu}\psi = T\frac{\partial \psi}{\partial x_\mu}$$

That is equal to the ordinary derivative transformed by T. The new derivative D_μ is denoted as a covariant derivative. The second order action in this case is:

$F^{(2)} : F_1 \rightarrow F_2$ where,

$$F_1 : \psi(x_\mu) \rightarrow \frac{\partial \psi(x_\mu)}{\partial x_\mu}$$

and $F_2 : T(x_\mu)\psi(x_\mu) \rightarrow T(x_\mu)\dfrac{\partial \psi(x_\mu)}{\partial x_\mu}$

For coherence we have:

$$F_2 T \psi = T \frac{\partial \psi}{\partial t}$$

$$F_2 T \psi = \frac{\partial}{\partial x_\mu}(T\psi) + [\, T \frac{\partial \psi}{\partial x_\mu} - \frac{\partial}{\partial x_\mu}(T\psi)]$$

$$= \frac{\partial}{\partial x_\mu}(T\psi) - \frac{\partial T}{\partial x_\mu}\psi$$

$$= D_\mu T \psi$$

Thus we have: $F_2 = D_\mu$.

We can prove that in mechanics the fundamental symmetry given by the least action for the Lagrangian L is still valid when we substitute the classical derivative with the covariant derivative D_μ where the classical derivative is compensated. The principle of the least action is:

$$\delta \int_{t_1}^{t_2} L(\psi(x,t), \frac{\partial \psi(x,t)}{\partial x_\mu}) dt dx = 0$$

We can prove that the least action is invariant when we substitute the ordinary derivative with the covariant derivative. So we have:

$$\delta \int_{t_1}^{t_2} L(\psi(x,t), \frac{\partial \psi(x,t)}{\partial x_\mu}) dt dx = \delta \int_{t_1}^{t_2} L(T\psi(x,t), D_\mu T\psi(x,t)) dt dx = 0$$

In classical mechanics the derivative is invariant. In field mechanics we have a local transformation T(x,t) which is the *reshape transformation* that is used in Jessel's work.

The covariant derivative D_μ, can be considered as differential reshape operator for the field $\psi(x,t)$. Here again we have another possible split in two different contexts, one before the differential reshape operation and the other after the differential reshape operation. The covariant derivative D_η before and after the transformation is unchanged. In the first context the derivative is $D_\eta \psi(x,t)$, in the second context the field is:

$$\psi'(x,t) = D_\mu \psi(x,t)$$

The result of the derivative in the second context is:

$$D_\eta \psi'(x,t) = D_\eta D_\mu \psi(x,t)$$

When we reshape the field $D_\eta \psi(x,t)$ by D_μ we obtain $D_\mu D_\eta \psi(x,t)$ for the second context. For the covariant defined by the function T(x,t) we have:

$$\frac{\partial^2 T}{\partial x_\mu \partial x_\eta} = \frac{\partial^2 T}{\partial x_\eta \partial x_\mu}$$

and,

$$(D_\eta D_\mu - D_\mu D_\eta) \psi(x,t) = 0$$

The derivative D_η in the first context is the same derivative as in the second context without any conflict. But when T is a path dependent integral such as:

$$T(x) = \int_x^B A_\mu dx^\mu$$

And here,

$$(D_\eta D_\mu - D_\mu D_\eta)\,\psi(x,t) \neq 0$$

where,

$$F_{\eta,\mu} = (D_\eta D_\mu - D_\mu D_\eta)\,\psi(x,t)$$

which is the Gauge field.

One example of a function that depends on the path is the phase at a point of the wave. When a wave starts from the point A and arrives at the point B, the phase in B depends on B and also on the previous path that follows between A and B. In this instance for the reshape of T for the phase in the wave function the reshape is dependent on the path followed.

9.6 Gravity Phenomena and Internal Adaptation

In the arena of the space-time every movement is free of force. We consider that the main symmetry in physics is represented by the inertial movement without any forces. We argue that the gravitational field is not really present but is only the compensating term for the asymmetry generated by the mass. Every trajectory in a gravitational field is embedded in a deformed space-time which changes the inertial movement. The equivalent principle, in which every body with any type of mass has the same acceleration in a gravitational field, is a local principle. It assures the inertial movement in a deformed geometric space-time context. The context changes in such a way as to maintain the inertial movement independent of the particular shape and mass of any type of the physical feature of the body. This is known as the equivalence principle.

When every point has an associated second order action which ensures that the inertia is constant when moving from one point to another under the gravitational field action, we can say that the gravitation field is only a compensating term in the second order action. This local invariance of the inertial movement means that physical objects inside a gravitational field have a fundamental symmetry. In the classical image we have the gravitational field in a flat space. In this way there is a conflicting situation between space and gravity. When we change the inertial law by a compensating term that generates the curvature of the space, the gravitational field disappears and we eliminate the conflicting situation.

9.6.1 Covariant Derivative D_k as Second Order Action

Given two contexts we have the coordinate x^i in one and the coordinate x'^i in the other.

In Figure 9.4 the two contexts are shown.

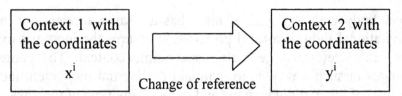

Figure 9.4. The two contexts and the two different references or coordinates.

In 8.2 and figure 8.1 we have shown the symmetry for the derivative operators in the adaptive field. We now use the same formal structure for the gravitational field.

Where,

$$H = F_k \frac{\partial^2 y^k}{\partial x^s \partial x^p}$$ was the adaptive field but now is the compensating

term for the gravitational field for the change of reference in space-time. We know that,

$$F_k \frac{\partial^2 y^k}{\partial x^s \partial x^p} = F_k \, \Gamma_{sp}{}^k$$

where $\Gamma_{sp}{}^k$ are the Christoffel symbols[13] that are not covariant when the reference is changed.

9.6.2 Extension of General Relativity and Quantum Mechanics

Given Einstein equation:

$$R_{i,j} = T_{i,j} - \frac{1}{2} g_{i,j} T$$

we can write,

$$D_k R_{i,j} - D_j R_{i,k} = D_k(T_{i,j} - \frac{1}{2}g_{i,j}T) - D_j(T_{i,k} - \frac{1}{2}g_{i,k}T)$$

It can be shown that,

$$R_{i,j} = [\, D_i \,, D_j \,]$$

we have:

$$D_k\,[D_j, D_i] - D_j\,[D_k, D_i] = D_k(T_{i,j} - \frac{1}{2}g_{i,j}T) - D_j(T_{i,k} - \frac{1}{2}g_{i,k}T)$$

Given the field ϕ_α we can write the previous equation as follows:

$$D_k\,[D_j, D_i]\,\phi_\alpha - D_j\,[D_k, D_i]\,\phi_\alpha = D_k(T_{i,j} - \frac{1}{2}g_{i,j}T)\,\phi_\alpha - D_j(T_{i,k} - \frac{1}{2}g_{i,k}T)\,\phi_\alpha$$

If in the ordinary gravitational field we assume that $D_k\,\phi_k = 0$, we have:

$$D_k\,[D_j, D_i]\,\phi_k - D_j\,[D_k, D_i]\,\phi_j = D_k\,[D_j, D_i]\,\phi_k - D_j\,[D_k, D_i]\,\phi_j - [D_j, D_i]\,D_k\phi_k + [D_k, D_i]\,D_j\,\phi_j$$

This can be written as follows:

$$D_k\,[D_i, D_j]\,\phi_\alpha - D_j\,[D_i, D_k]\,\phi_\alpha = [D_i, [D_j, D_k]\,\phi_\alpha$$

In conclusion

$$[D_i, [D_j, D_k]]\,\phi_\alpha = D_k(T_{i,j} - \frac{1}{2}g_{i,j}T)\,\phi_\alpha - D_j(T_{i,k} - \frac{1}{2}g_{i,k}T)\,\phi_\alpha$$

However because $T = g^{ij} T_{ij}$ and $D_k\,g^{ij} = 0$, we have:

$$D_k g_{i,j} T - D_j g_{i,k} T = D_k g_{i,j} g^{i,j} T_{i,j} - D_j g_{i,k} g^{i,k} T_{i,k} = D_k T_{i,j} - D_j T_{i,k}$$

and,

$$[D_i,[D_j,D_k]]\,\phi_\alpha = \frac{\phi_\alpha}{2}[\,D_k T_{i,j} - D_j T_{i,k}\,] = \frac{\phi_\alpha}{2}\,\mathrm{Rot}T_{\alpha,\beta} = \chi\,J_{k,i,j}$$

$$(9.2)$$

Where,

$J_{k,i,j}$ = gravitational current..

$T_{\alpha\beta}$ = Tensor Energy-momentum

$T^{00} = \mu\,c^2$ Density of energy , where μ is the density of matter

$T^{0i} = c\,\mu\,v^i$ Density of momentum

$T^{ji} = \mu\,v^i\,v^j$ Density of the Kinematics energy

$$D_q\,T^{pq} = T^{pq}{}_{/q} = \frac{\partial T^{pq}}{\partial x^q} - \frac{1}{2}\Gamma^{rs}\frac{\partial g_{rs}}{\partial x^i} = 0$$

$$(9.3)$$

In (9.3) the covariant derivative of the Energy Momentum Tensor is obtained by the tensor calculus. For the physical symmetry or invariance of the Energy Momentum Tensor the covariant derivative must be equal to zero.

We can , by an integration process , obtain conservation of the total energy momentum before using the general relativity. Now the matter and gravity can be transformed into energy and momentum and vice versa. So we have a more complex invariant of energy , momentum and matter together. The 9.3 is the continuity equation and an extension of the Conservation of Energy and Momentum. For the general property of the divergence operator in space-time:

Div Rot $T_{\alpha\beta} = 0$,

and for (9.2) we obtain:

Div $J_{k,i,j} = 0$.

And the gravitational current $J_{k,i}$, is invariant.

When $D_k \phi_k$ is different from zero, we can write the previous equation as follows:

$$D_i[D_j, D_k]\phi\alpha = \frac{\phi\alpha}{2} RotT_{\alpha,\beta} + [D_j, D_k] D_i\phi\alpha$$

or,

$$D_i R_{k,j}\phi\alpha = \frac{\phi\alpha}{2} RotT_{\alpha,\beta} + R_{k,j} D_i\phi\alpha$$

When we introduce a *quantum gravitational current* $QJ_{i,k,j} = R_{k,j} D_i \phi\alpha$ we have:

$$D_i R_{k,j}\phi\alpha = \frac{\phi\alpha}{2} J_{i,j,k} + QJ_{i,j,k}$$

In a vacuum,

$$J_{i,jk} = 0,$$

and $D_i R_{k,j}\phi\alpha = QJ_{i,j,k}$.

Empty space has a curvature which is not zero. The vacuum has a physical effect and generates gravitation. We can say that because the equivalent principle is true only in the vacuum the curvature of the space is zero and the equivalent principle is, under these conditions, false.

When $D_\gamma\phi$ is different from zero the field equation is:

$$D_h R^a_{ijk} D_a\phi = \chi J_{hjk} D_a\phi - R^a_{hjk} D_a D_a\phi$$

The new source of gravitational field is:

$$\Lambda_{hij} = -R^a_{hjk} \frac{D_a^2 \phi}{D_a \phi} \tag{9.4}$$

and is a negative source comparable with the cosmological constant. In the case of the negative gravitational field the gravitational force is repulsive as recent experiments have shown [27, 28].

The cosmological constant in (9.4) is proportional to the second derivative of the quantum scalar field ϕ and is also proportional to the curvature of the space-time given by the tensor R^a_{hjk}.

For the Kline Gordon equation we have:

$\eta^{\alpha\beta} \partial_\alpha \phi \partial_\beta \phi = m^2 \phi$ and in the general coordinates $\eta^{\alpha\beta} D_\alpha \phi D_\beta \phi = m^2 \phi$

The cosmological constant can be written as:

$$\Lambda_{hij} = -R^a_{hjk} \frac{m^2 \phi}{\left(\dfrac{\partial \phi}{\partial x_a} + \Gamma^\alpha_{\beta\gamma} \dfrac{\partial \phi}{\partial x_\alpha}\right)}$$

For weak and nearly static gravitational fields we have:

$$R_{44} = -\frac{\Delta G}{c^2} \text{ and } \Gamma_{\alpha,44} = \frac{1}{2}\left(\frac{\partial g_{44}}{\partial x_\alpha}\right) \text{ with } g_{44} = 1 - \frac{G}{c^2}$$

and the intensity Λ of the cosmological constant Λ_{hij} is:

$$\Lambda = \frac{m^2 \phi \Delta G}{c^2 + \dfrac{\partial G}{\partial x_\mu} \dfrac{\partial \phi}{\partial x_\eta}} = \frac{m^2 \phi \Delta G}{c^2} = \frac{S_Q S_G}{c^2}$$

Where G is the gravitational field, $S_Q = m^2 \phi$ is the source of the quantum waves and S_G are the sources of the scalar gravitational field. Because,

$S_G \approx G_N \rho$
$= 6.673 \ 10^{-11} \ m^3 \ kg^{-1}s^{-2} \ 3 \ 10^{-22} \ kg \ m^{-3}$
$= 1.8 \ 10^{-32} \ , 1/c^2 \approx \ 10^{-17} \ s^2 \ m^{-2}$

Where ρ is the density of local halo, we have,

$\Lambda \approx 10^{-49} \ [m^2 \ \phi \ ]$ that is comparable with the experimental value of $\Lambda \approx 2.853 \ 10^{-51} \ h_0^{-2} \ m^2$

where $h_0 \approx 0.71$ is a scalar parameter [31]. The cosmological constant is a physical effect of the interaction between the gravitational field and the scalar quantum field. Using the scalar quantum field we can know the probability of detecting a particle in a point of the space-time.

In the paper "Coherence effects in neutron diffraction and gravity experiments" [30] it is shown that gravity has quantum effects on the phase of the wave function. Here we propose another phenomenon. That is a rapid change of the wave function in quantum mechanics can generate negative sources for gravity or antigravity as shown in (9.4).

9.7 Physical Internal Adaptive Field which Transforms Space with Curvature into Flat Space

Given the coordinate transformation:

$$y_1 = x_1 \cos(x_2) = R \cos(\theta)$$

$$y_2 = x_1 \sin(x_2) = R \sin(\theta)$$

The affine transformation $\dfrac{\partial y^{\mu}}{\partial x^{\nu}}$ is:

$$
\begin{bmatrix}
\dfrac{\partial y_1}{\partial x_1} & \dfrac{\partial y_1}{\partial x_2} \\[2mm]
\dfrac{\partial y_2}{\partial x_1} & \dfrac{\partial y_2}{\partial x_2}
\end{bmatrix}
=
\begin{bmatrix}
\cos(x_2) & -x_1 \sin(x_2) \\
\sin(x_2) & x_1 \cos(x_2)
\end{bmatrix}
\tag{9.5}
$$

The (9.5) can be written (Table 9.1) in the similar way as the tables (objects, features) shown in chapter 5.

Table 9.1. Features and Objects in affine transformation of the coordinates (x_1, x_2) in (9.5).

Objects	Coordinate x_1 Feature 1	Coordinate x_2 Feature 2
Object 1	$\cos(x_2)$	$-x_1 \sin(x_2)$
Object 2	$\sin(x_2)$	$x_1 \cos(x_2)$

Here the sample of the features and objects are functions of x_1 and x_2. With respect to the tables in chapter 5 we have an extension from simple real numbers to functions.

In chapter 5 using the table (objects and features) we calculate the tensor g. In a like manner we are able to calculate G for Table 9.1 using the relation (9.5):

$$
g = \begin{bmatrix} \dfrac{\partial y_1}{\partial x_1}\dfrac{\partial y_1}{\partial x_1} + \dfrac{\partial y_2}{\partial x_1}\dfrac{\partial y_2}{\partial x_1} & \dfrac{\partial y_1}{\partial x_1}\dfrac{\partial y_1}{\partial x_2} + \dfrac{\partial y_2}{\partial x_1}\dfrac{\partial y_2}{\partial x_2} \\[2mm] \dfrac{\partial y_1}{\partial x_2}\dfrac{\partial y_1}{\partial x_1} + \dfrac{\partial y_2}{\partial x_2}\dfrac{\partial y_2}{\partial x_1} & \dfrac{\partial y_1}{\partial x_2}\dfrac{\partial y_1}{\partial x_2} + \dfrac{\partial y_2}{\partial x_2}\dfrac{\partial y_2}{\partial x_2} \end{bmatrix} = \begin{bmatrix} 1 & 0 \\ 0 & x_1^{\,2} \end{bmatrix} \tag{9.6}
$$

In chapter 5 with g we have written the quadratic form. Now with the g in (9.6) we obtain:

$$
ds^2 = dx_1^{\,2} + x_1^{\,2}\, dx_2^{\,2} \tag{9.7}
$$

Here we have not the ordinary quadratic form in the Euclidean space or $ds^2 = dx_1^{\,2} + dx_2^{\,2}$. The space of the feature is not a Euclidean space. In (9.7) x_1 is present in the coefficients of the quadratic form. This will generate a non linear quadratic form. In chapter 5 we generated special transformations of the coordinates by which we can transform a general quadratic form into a quadratic form the coefficients of which are real numbers.

In Chapter 5 we created the transformation T for which:

$$
x_j^* = T\, x_j
$$

with the property that:

$$
\Sigma\, g_{ij}\, x_i\, x_j = \Sigma\, (\, x_i^* \,)^2
$$

In this chapter, because we calculate the space of the feature using the partial derivative and the coefficients of the quadratic form are not simple real numbers, we change the space of the feature in a different way.

In the paper of Ning Wu [29] a new definition of the derivate is given as:

$$D_\mu = \partial_\mu - gC_\mu{}^\nu \partial_\nu \qquad (9.8)$$

In a two dimensional form we can write (9.8) as:

$$\begin{bmatrix} D_1 \\ D_2 \end{bmatrix} = \begin{bmatrix} 1-gC_1^1 & -C_1^2 \\ -C_2^1 & 1-gC_2^2 \end{bmatrix} \begin{bmatrix} \partial_1 \\ \partial_2 \end{bmatrix} \qquad (9.9)$$

Using the new derivative we can rewrite the affine transformation as:

$$\begin{bmatrix} D_1y_1 & D_2y_1 \\ D_1y_2 & D_2y_2 \end{bmatrix} = \begin{bmatrix} \dfrac{\partial y_1}{\partial x_1}(1-gC_1^1) - \dfrac{\partial y_1}{\partial x_2}gC_1^2 & -\dfrac{\partial y_1}{\partial x_1}gC_2^1 + \dfrac{\partial y_1}{\partial x_2}(1-gC_2^2) \\ \dfrac{\partial y_2}{\partial x_1}(1-gC_1^1) - \dfrac{\partial y_2}{\partial x_2}gC_1^2 & -\dfrac{\partial y_2}{\partial x_1}gC_2^1 + \dfrac{\partial y_2}{\partial x_2}(1-gC_2^2) \end{bmatrix}$$

Or using Equation (9.5) we have:

$$\begin{bmatrix} D_1y_1 & D_2y_1 \\ D_1y_2 & D_2y_2 \end{bmatrix} = \begin{bmatrix} (1-gC_1^1)\cos(x_2)+gC_1^2x_1\sin(x_2) & -gC_2^1\cos(x_2)-(1-gC_2^2)x_1\sin(x_2) \\ (1-gC_1^1)\sin(x_2)-x_1gC_1^2\cos(x_2) & -gC_2^1\sin(x_2)+(1-gC_2^2)x_1\cos(x_2) \end{bmatrix}$$

$$(9.10)$$

where g is a parameter.

$\begin{bmatrix} C_1^1 & C_1^2 \\ C_2^1 & C_2^2 \end{bmatrix}$ are the components of the internal adaptive physical field.

Using the matrix (9.10) we obtain the new Table 9.2 of the objects and the features:

Table 9.2. Features and objects with internal adaptive field C.

Objects	Coordinate x_1, Feature 1	Coordinate x_2, Feature 2
Object 1	$(1 - gC_1^1)\cos(x_2)$ $+ gC_1^2 x_1 \sin(x_2)$	$(1 - gC_2^2)\, x_1 \sin(x_2)$ $- gC_2^1 \cos(x_2)$
Object 2	$(1 - gC_1^1)\sin(x_2)$ $- gC_1^2 x_1 \cos(x_2)$	$(1 - gC_2^2)\, x_1 \sin(x_2)$ $- gC_2^1 \cos(x_2)$

With the new Table 9.2 of objects and features we have the tensor g*:

$$g^* = \begin{bmatrix} D_1 y_1 D_1 y_1 + D_1 y_2 D_1 y_2 & D_1 y_1 D_2 y_1 + D_1 y_2 D_2 y_2 \\ D_2 y_1 D_1 y_1 + D_2 y_2 D_1 y_2 & D_2 y_1 D_2 y_1 + D_2 y_2 D_2 y_2 \end{bmatrix}$$

$$= \begin{bmatrix} x_1^2 (gC_1^2)^2 + (gC_1^1 - 1)^2 & gC_1^2 (gC_2^2 - 1)x_1^2 + gC_2^1 (1 - gC_1^1) \\ gC_1^2 (gC_2^2 - 1)x_1^2 + gC_2^1 (1 - gC_1^1) & x_1^2 (gC_2^2 - 1)^2 + (gC_2^1)^2 \end{bmatrix}$$

We then have the quadratic form:

$$s^2 = [x_1^2 (gC_1^2)^2 + (gC_1^1 - 1)^2]\, dx_1^2 + 2\,[\, gC_1^2 (gC_2^2 - 1)x_1^2 + gC_2^1 (1 - gC_1^1)]\, dx_1 dx_2 + [\, x_1^2 (gC_2^2 - 1)^2 + (gC_2^1)^2]\, dx_2^2 \qquad (9.11)$$

We can then compute an internal adaptive physical field in such a way that the (9.11) is converted to the quadratic form $s^2 = dx_1^2 + dx_2^2$. For:

$$\begin{bmatrix} gC_1^{\,1} & gC_1^{\,2} \\ gC_2^{\,1} & gC_2^{\,2} \end{bmatrix} = \begin{bmatrix} \left\{ \dfrac{1+\sqrt{1-\Omega^2}}{1-\sqrt{1-\Omega^2}} \right\} & -\dfrac{\Omega}{R} \\[2em] \Omega & \left\{ \begin{aligned} 1+\dfrac{\sqrt{1-\Omega^2}}{R} \\ 1-\dfrac{\sqrt{1-\Omega^2}}{R} \end{aligned} \right\} \end{bmatrix}$$

Where Ω is an arbitrary parameter we have:

$$\begin{bmatrix} D_1 \\ D_2 \end{bmatrix} = \begin{bmatrix} \left\{ \dfrac{-\sqrt{1-\Omega^2}}{\sqrt{1-\Omega^2}} \right\} & \dfrac{\Omega}{R} \\[2em] -\Omega & \left\{ \begin{aligned} -\dfrac{\sqrt{1-\Omega^2}}{R} \\ \dfrac{\sqrt{1-\Omega^2}}{R} \end{aligned} \right\} \end{bmatrix} \begin{bmatrix} \partial_1 \\ \partial_2 \end{bmatrix} \qquad (9.12)$$

and the tensor g* is:

$$g^* = \begin{bmatrix} 1 & 0 \\ 0 & 1 \end{bmatrix} \qquad (9.13)$$

It is noted that in the computation of g* the arbitrary parameter Ω will disappear. We conclude that for every arbitrary value of Ω and for (9.13) we can obtain a flat space with the following quadratic form

$$s^2 = dx_1^{\,2} + dx_2^{\,2} \qquad (9.14)$$

Using the new derivatives, we can calculate the expression:

$$F = (D_1 D_2 - D_2 D_1) \phi = [D_1 , D_2] \phi$$

And Equation (9.12) enables us to calculate F to obtain:

$$F_{1,2} = [D_1, D_2] \phi(R, \theta) = \frac{1}{g} \frac{1}{R^2} [-\frac{\partial \phi(R, \theta)}{\partial \theta}] \qquad (9.15)$$

In the computation of $F_{1,2}$ the parameter Ω disappears. The parameter Ω is a Gauge Parameter inside the potential $C_\mu{}^\nu$ but is not present in the final Euclidean form (9.14) and the function F.

When $M = -\dfrac{\partial \phi(R, \theta)}{\partial \theta}$ we have the Newton gravitational field:

$$F = G \frac{M}{R^2} \qquad (9.16)$$

Equation (9.16) clarifies the meaning of the variable F. The variable F is the gravitational field when $G = 1/g$ is the gravitational constant We may consider ϕ to be the *mass potential*, and every variation of this potential will result in the generation of real mass.

9.8 Adaptive Current in the Second Order Action and Physical Action

Given a one-dimensional system of N particle points, each of mass m , coupled elastically by a series of springs of length a, each with a constant k force and η_i the displacement of the particle i from its equilibrium position, the equation of motion of the system may be obtained by setting the variation of the action equal to zero.

$$\delta A = \delta \int_{t_1}^{t_2} L(\eta_i, \frac{d\eta_i}{dt})dt = 0$$

where,

$$L = \frac{1}{2}\sum_{i=1}^{N}[m(\frac{d\eta_i}{dt})^2 - k(\eta_{i+1} - \eta_i)^2]$$

$$= \frac{1}{2}\sum_{i=1}^{N}[\frac{m}{a}(\frac{d\eta_i}{dt})^2 - ka(\frac{\eta_{i+1} - \eta_i}{a})^2]a \to \frac{1}{2}\sum_{i=1}^{N}[\rho(\frac{d\eta_i}{dt})^2 - Y(\frac{\partial\eta}{\partial x})^2]dx$$

Then,

$$\delta A = \delta \int_{t_1}^{t_2} L(\eta_i, \frac{d\eta_i}{dt})dt$$

$$= \delta \int_{t_1}^{t_2}\int_{x_1}^{x_2}[\rho(\frac{\partial\eta}{\partial t})^2 - Y(\frac{\partial\eta}{\partial x})^2]dtdx = 0$$

is in the space of the mechanical phase of a field in a space of infinite dimensions.

From the Noether's theorem, we have:

$$\delta A = \delta \int_{\Omega} L[\eta(x_\lambda), \partial_\lambda\eta(x_\lambda)]\, d^4x$$

$$= \delta \int_{\Omega'} L[\eta'(x'_\lambda), \partial_\lambda\eta'(x'_\lambda)]\, d^4x' = 0$$

In Figure 9.5 we show the two contexts:

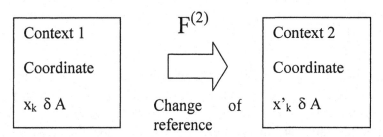

Figure 9.5. The two contexts where the variation of the mechanical action
A is identical for both contexts.

The current is:

$$J_\mu = [-\frac{\partial L}{\partial(\partial_\mu \eta_\sigma)} \partial^\lambda \eta^\sigma + \delta_\alpha^\mu L] \, \delta x_\lambda + \frac{\partial L}{\partial(\partial_\mu \eta_\sigma)} \delta \eta^\sigma$$

For the Conservation of Energy and Momentum in the field we use
the translation transformation of the space-time.

$$J_\mu = [-\frac{\partial L}{\partial(\partial_\mu \eta_\sigma)} \partial^\lambda \eta^\sigma + \delta_\alpha^\mu L] \, \delta x_\lambda$$

and,

$$\frac{\partial J_\mu}{\partial x_\mu} = 0$$

That is the current density of the field is conserved. The conserved
inertial current for a scalar field corresponding to the Global Gravitational Gauge Symmetry of Ning Wu is:

$$J_{i\alpha}^\mu = e^{I(C)} (-\frac{\partial L_0}{\partial(\partial_\mu \phi)} \partial_\alpha \phi - \frac{\partial L_0}{\partial(\partial_\mu C_v^\beta)} \partial_\alpha C_v^\beta + \delta_\alpha^\mu L_0)$$

where,

$$L_0 = -\frac{1}{2}\eta^{\mu\nu}D_\mu\phi\,D_\nu\phi - \frac{m^2}{2}\phi^2 - \frac{1}{4}\eta^{\mu\rho}\eta^{\nu\sigma}\eta_{2\alpha\beta}F^\alpha_{\mu\nu}F^\beta_{\rho\sigma} \tag{9.17}$$

Where the adaptive field C is a variable in the space where the adaptive current is calculated. For the definition of the tensor C, for the gravitational field potential in (9.8) and for equation (9.17), the dynamical equation in quantum mechanics for the gravitational field is defined. We consider that the image in the curvature space in chapter 9.6 and in the flat space in chapter 9.7, gives two complementary images of the quantum interpretation of the gravitational field. They provide the possibility of future unification between the gravitational field and all the other fields.

9.9 Internal Adaptive Entity Caused by the Division of a Physical System into Several Parts

Consider the physical system shown in Figure 9.6 where a sound wave moves from the left to the right with different velocities c_1 and c_2 in each part.

In Figure 9.6 the sound waves enter one tube with velocity c_1 and go out of the other tube with velocity c_2. The propagation of the wave of sound pressure $n(x, t)$ in a tube or in the system may be represented by a differential model of the form:

$$(\frac{\partial^2}{\partial x^2} - \frac{1}{c(x)^2}\frac{\partial^2}{\partial t^2})\,n(x,t) = P\,n(x,t) = S \tag{9.18}$$

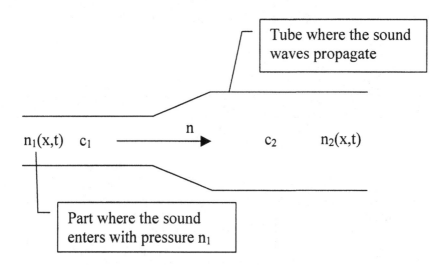

Figure 9.6. A Physical System composed of two parts: in the part on the left the sound moves at a constant velocity c_1. In the part on the right the sound moves at constant velocity c_2. In such a physical system the velocity of the sound is not constant but can be divided into two subsystems where the velocity is constant.

where S is the sound source. Because the velocity c(x) is a function of the position x in the tube, the solution of the differential equation is difficult.. When the velocity is constant we can obtain the solution of the differential equation. When we divide the system into two subsystems, each of them having constant velocity, we have:

$$n_1(x,t) = \mu_1\, n(x,t) \,,\; n_2(x,t) = \mu_2\, n(x,t)$$

For the two variables we obtain the partial equations:

$$(\frac{\partial^2}{\partial x^2} - \frac{1}{c_1^2}\frac{\partial^2}{\partial t^2})\mu_1 n(x,t) = P_1\mu_1 n(x,t) = P_1 n_1(x,t) = \mu_1 S + S_1^{ex} = S_1 + S_1^{ex}$$

$$(\frac{\partial^2}{\partial x^2} - \frac{1}{c_2^2}\frac{\partial^2}{\partial t^2})\mu_2 n(x,t) = P_2\mu_2 n(x,t) = P_2 n_2(x,t) = \mu_2 S + S_2^{ex} = S_2 + S_2^{ex}$$

Because,

$\mu_1 + \mu_2 = 1$, we have $\mu_2 = 1 - \mu_1$ and because $S_1^{ex} + S_2^{ex} = 0$, $S_2^{ex} = -S_1^{ex}$

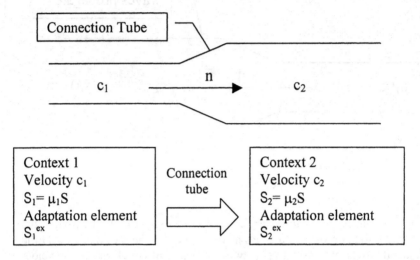

Figure 9.7. In this figure we show the two contexts related to the two velocities c_1 and c_2.

We have now:

$P_1 \mu_1 n (x,t) + P_2 (1-\mu_1) n (x,t) = P n(x,t)$

And,

$$(\frac{\partial^2}{\partial x^2} - \frac{1}{c_1^2}\frac{\partial^2}{\partial t^2})\mu_1(x)n(x,t) + (\frac{\partial^2}{\partial x^2} - \frac{1}{c_2^2}\frac{\partial^2}{\partial t^2})(1-\mu_1(x))n(x,t) = (\frac{\partial^2}{\partial x^2} - \frac{1}{c(x)^2}\frac{\partial^2}{\partial t^2})n(x,t)$$

Thus:

$$(\frac{\partial^2 \mu_1(x)n(x,t)}{\partial x^2} - \frac{\mu_1(x)}{c_1^2}\frac{\partial^2 n(x,t)}{\partial t^2}) +$$

$$(\frac{\partial^2 (n(x,t) - \mu_1(x)n(x,t))}{\partial x^2} - \frac{1 - \mu_1(x)}{c_2^2}\frac{\partial^2 n(x,t)}{\partial t^2})$$

$$= (\frac{\partial^2 n(x,t)}{\partial x^2} - \frac{1}{c(x)^2}\frac{\partial^2 n(x,t)}{\partial t^2})$$

Resulting in:

$$(-\frac{\mu_1(x)}{c_1^2}\frac{\partial^2 n(x,t)}{\partial t^2}) + (-\frac{1 - \mu_1(x)}{c_2^2}\frac{\partial^2 n(x,t)}{\partial t^2})$$

$$= (-\frac{1}{c(x)^2}\frac{\partial^2 n(x,t)}{\partial t^2})$$

And,

$$\frac{\mu_1(x)}{c_1^2} + \frac{1 - \mu_1(x)}{c_2^2} = \frac{1}{c(x)^2}$$

In conclusion we have,

$$\mu_1(x) = \frac{c_1^2}{c(x)^2}(\frac{c_2^2 - c(x)^2}{c_2^2 - c_1^2}) \qquad (9.19)$$

When,

$$c(x) = \frac{c_1 - c_2}{1 + e^{-\alpha x}} + c_1$$

For $c_1 = 4$ and $c_2 = 5$ and $\alpha = 1$ we have Figure 9.8 with arbitrary units.

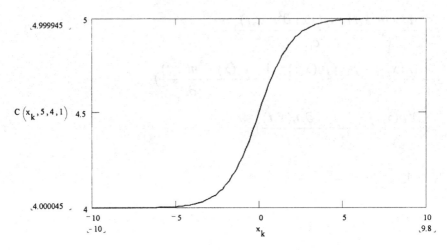

Figure 9.8. Variation of the velocity inside the tube in the Figure 9.5.

Using the formula (9.19) we obtain the graph shown in Figure 9.9 with arbitrary units.

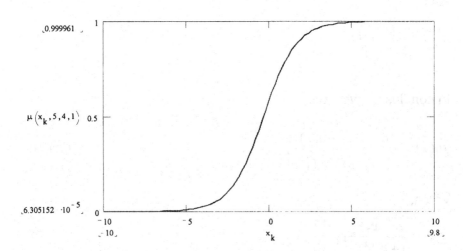

Figure 9.9. Variation of the Parameter μ (x).

When we calculate the solution of the equation in a vacuum:

$$n(x,t) = u(x)\exp(-1\,\omega\,t)$$

where $\omega = 2\pi f$ and f is the frequency of the wave.

Since,

$$S_2^{ex} = -S_1^{ex}$$

we have:

$$F(x) = \frac{ik_1}{2}\int e^{-ik_1|x-x'|}[\,\mu_1(x')S(x')dx' + \frac{ik_2}{2}\int e^{-ik_2|x-x'|}[\,\mu_2(x')S(x')dx'$$

$$[(\frac{d^2}{dx^2}+k_1^2)\mu_1(x)-\mu_1(x)(\frac{d^2}{dx^2}+k^2)][G\ (c_1)*(S_1+S_1^{ex})+G\ (c_2)*(S_2-S_1^{ex})] = S_1^{ex}$$

and,

$$[\frac{d^2\mu_1(x)}{dx^2}-\mu_1(x)F(x)\,]-\mu_1(x)[\ \frac{ik_1}{2}\int e^{-ik_1|x-x'|}S_1^{ex}dx'-\frac{ik_2}{2}\int e^{-ik_2|x-x'|}\ S_1^{ex}dx'\ ] = S_1^{ex}$$

Where the green function $G(c)$ is:

$$G(c) = \frac{1}{2ik}\exp\ (-ik\,|\,x-x'|)$$

and $k = \dfrac{\omega}{c(x)}$, $k_1 = \dfrac{\omega}{c_1}$, $k_2 = \dfrac{\omega}{c_2}$

The product " * " is the convolution product.

$$u(x) = G\ (c)*S = \int G\ (x,x')S(x')dx'$$

The convolution product is the superposition of the fields (Green function) and results from the elementary sources within S.

Bibliography

1. Penna M.P., Pessa E. and Resconi G., General System Logical Theory and its role in cognitive psychology, Third European Congress on Systems Science, Rome, 1-4 October, 1996.

2. Tzvetkova G.V. and Resconi G., Network recursive structure of robot dynamics based on GSLT, European Congress on Systems Science, Rome, 1-4 October, 1996.

3. Resconi G. and Tzvetkova G.V., Simulation of Dynamic Behaviour of robot manipulators by General System Logical Theory,14-th International Symposium "Manufacturing and Robots" 25-27 June, Lugano Switzeland, pp.103-106, 1991.

4. Resconi G. and Hill G., The Language of General Systems Logical Theory: a Categorical View, European Congress on Systems Science, Rome, 1-4 October 1996.

5. Rattray C., Resconi G. and Hill G., GSLT and software Development Process, Eleventh International Conference on Mathematical and Computer Modelling and Scientific Computing, Georgetown University, Washington D.C.March 31-April 3,1997

6. Mignani R., Pessa E. and Resconi G., Commutative diagrams and tensor calculus in Riemann spaces, Il Nuovo Cimento 108B(12) December 1993.

7. Petrov A.A., Resconi G., Faglia R. and Magnani P.L., General System Logical Theory and its applications to task description for intelligent robot. In Proceeding of the sixth International Conference on Artificial Intelligence and Information Control System of Robots, Smolenize Castle,Slovakia, September 1994.

8. Minati G. and Resconi G., Detecting Meaning, European Congress on Systems Science, Rome, 1-4 October 1996.

9. Kazakov G.A. and Resconi G., Influenced Markovian Checking Processes By General System Logical Theory, International Journal General System, vol.22, pp.277-296, 1994.

10. Saunders Mac Lane, *Categories for Working Mathematician*, Springer-New York, Heidelberg Berlino, 1971.

11. Wymore A.W., *Model-Based Systems Engineering*. CRC Press, 1993.

12. Mesarovich M.D. and Takahara Y., *Foundation of General System Theory*. Academic Press, 1975.

13. Mesarovich M.D. and Takahara Y., Abstract System Theory, *In* Lecture Notes in Control and Information Science *116*. Springer Verlag, 1989.

14. Resconi G. and Jessel M., A General System Logical Theory, International Journal of General Systems, vol.12, pp.159-182, 1986.

15. Resconi G. and Wymore A.W., Tricotyledon Theory of System amd General System Logical Theory, Eurocast'97, 1997.

16. Fatmi H.A., Marcer P.J., Jessel M. and Resconi G., Theory of Cybernetics and Intelligent Machine based on Lie Commutators, vol.16, no.2, pp123-164, 1990.

17. Kalman R.E., Falb P.L. and Arbib M.A., *Topics in Mathematical System Theory*, McGraw-Hill Publ, 1969.

18. Mesarovic M.D. and Takahara Y., 1989 Abstract Systems Theory, Lecture Notes in Control and Information Systems, Springer-Langer, 1989.

19. Padulo L. and Arbib M.A., *System Theory*, WB Saunders, 1974.

20. Resconi G., Rattray C. and Hill G., The Language of General Systems Logical Theory (GSLT), International Journal of General Systems, vol.28, no.4-5, pp.383-416, 1999.

21. Lin Y., Development of New Theory with Generality to unify diverse disciplines of knowledge and capability of applications, Int. J. General System, vol.23, pp.221-239, 1995.

22. Gurevich Y., Sequential Abstract State Machines Capture Sequential Algorithms, ACM Transactions on Computational Logic vol.1, no.1, pp.77-111, July 2000.

23. Resconi G. and Jessel M., A General System Logical Theory, International Journal of General Systems, vol.12, pp.159-182, 1986.

24. Auger and Resconi G., Hilbert Space and Dynamical Hierarchical Systems, Intern. J. General Systems, vol.16, pp.235-252, 1990.

25. Jessel M., Une Methode pour etudier certains problemes d'interaction, Le journal de Physique et le radium tome 17, pp.1022, December 1956.

26. Resconi G., Rattray C. and Hill G., The language of general systems logical theory (GSLT), Int.J.General Systems, vol.28, no.4-5, pp.383-416, 1999.

27. Burrows A., Supernova explosions in the Universe, Nature, vol. 403, pp.727-733, 17 February 2000.

28. Zehavi Idit, Dekel Avishai, Evidence for a positive cosmological constant from flows of galaxies and distant supernovae, Nature, vol.401, pp.252-254, 16 September 1999.

29. Wu N., Gauge Theory of Gravity, preprint arXiv: hep-th/0109145 v3, 21 Sep.2001.

30. Daniel M.Greenberger and Overhauser A.W., Reviews of Modern Physics, vol.51, no.1, January 1979.

31. The European Physical Journal C Review of Particle physics, vol.15, no.1-4. 2000.

Chapter Ten

Practical Applications of Agents in Robots and Evolution Population

10.1 Invariant Properties and Symmetries for a Two-Dimensional Robot

In the robot SCARA [John J.Craig, 1955] [1] the equation that connects the work space (x,y) with the joint space (α,β) is:

$$\begin{cases} X = \psi_1 = \cos(\alpha) + \cos(\alpha + \beta) \\ Y = \psi_2 = \sin(\alpha) + \sin(\alpha + \beta) \end{cases} \tag{10.1}$$

These positions are represented in Figure.10.1 where we show the mechanical system of the robot SCARA.

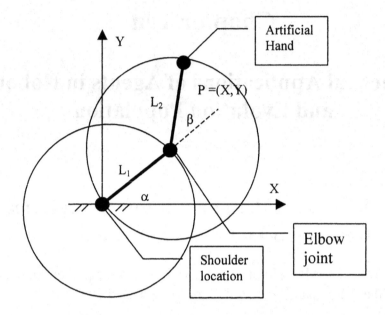

Figure 10.1. Simple Image of the Robot SCARA. In the coordinates X and Y the arm of the robot of length L_1 and L_2 of its two parts is shown. At the end of the arm there is an artificial hand able to grasp objects.

The expression (10.1) can be written as a table containing the two features as follows:

Table 10.1. Expression of the Equation (10.1) as a table of Objects and Features.

α	$X = \psi_1$	$Y = \psi_2$
α_1	$\cos(\alpha_1) + \cos(\alpha_1 + \beta)$	$\sin(\alpha_2) + \sin(\alpha_2 + \beta)$
α_2	$\cos(\alpha_2) + \cos(\alpha_2 + \beta)$	$\sin(\alpha_2) + \sin(\alpha_2 + \beta)$
..		
α_n	$\cos(\alpha_n) + \cos(\alpha_n + \beta)$	$\sin(\alpha_n) + \sin(\alpha_n + \beta)$

As in Chapter 5, which also has tables of the objects and features, we can also use the data in Table 10.1 to calculate the tensor g (β). We obtain:

$$g(\beta) = \begin{bmatrix} \sum_{k=1}^{n} X_k^2 & \sum_{k=1}^{n} X_k Y_k \\ \sum_{k=1}^{n} X_k Y_k & \sum_{k=1}^{n} Y_k^2 \end{bmatrix}$$

$$= \begin{bmatrix} \sum_{k=1}^{n} [\cos(\alpha_k)+\cos(\alpha_k+\beta)]^2 & \sum_{k=1}^{n} [\cos(\alpha_k)+\cos(\alpha_k+\beta)][\sin(\alpha_k)+\sin(\alpha_k+\beta)] \\ \sum_{k=1}^{n} [\cos(\alpha_k)+\cos(\alpha_k+\beta)][\sin(\alpha_k)+\sin(\alpha_k+\beta)] & \sum_{k=1}^{n} [\sin(\alpha_k)+\sin(\alpha_k+\beta)]^2 \end{bmatrix}$$

When $n \to \infty$ we have $\Sigma \to \int$ and the tensor g (β) then becomes:

$$g(\beta) = \begin{bmatrix} \int_{\Omega} [\cos(\alpha)+\cos(\alpha+\beta)]^2 d\alpha & \int_{\Omega} [\cos(\alpha)+\cos(\alpha+\beta)][\sin(\alpha)+\sin(\alpha+\beta)]d\alpha \\ \int_{\Omega} [\cos(\alpha)+\cos(\alpha+\beta)][\sin(\alpha)+\sin(\alpha+\beta)]d\alpha & \int_{\Omega} [\sin(\alpha)+\sin(\alpha+\beta)]^2 d\alpha \end{bmatrix}$$

With the formal integration where $\Omega = [\,0\,,\,\pi\,]$, we have:

$$g(\beta) = \begin{bmatrix} \pi(1+\cos(\beta)) & 0 \\ 0 & \pi(1+\cos(\beta)) \end{bmatrix}$$

and for the quadratic form:

$$D(\beta) = [\pi(1+\cos(\beta))]\,[\psi_1^2 + \psi_2^2]$$

which is invariant when the angle α is changed. In chapter 5 we used the *quadratic form as an invariant when moving from one context to another.* For the previous quadratic form $S = \psi_1^2 + \psi_2^2$ does not depend on the angle α. We have:

$$[\,\psi_1\,]^2 + [\,\psi_2\,]^2 = 2\pi(1+\cos(\beta)) \tag{10.2}$$

The angle β can now be calculated by using (10.2) when the values of X and Y in (10.1) are known. When the angle β is known from (10.2) the angle α may be calculated by using (10.1). From the reference X, Y in the work space, we compute the joint space of the angles (α, β) for the control of the robot. By this means the inverse kinematics problem in robot SCARA is resolved.

Because D is invariant when we change the angle α, we may deduce that S also does not change when the angle α changes. The second order action in Figure 10.2 has symmetry.

For the second order when the angle α is changed, S does not change.

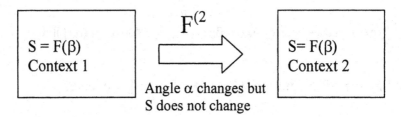

Figure 10.2. When the angle α is altered the context is also changed. The rule F in the first context is the same as that in the second context.

10.1.1 Invariant Properties and Symmetries for a Three - Dimensional Robot

Figure 10.3 shows a robot with an arm in three dimensions.

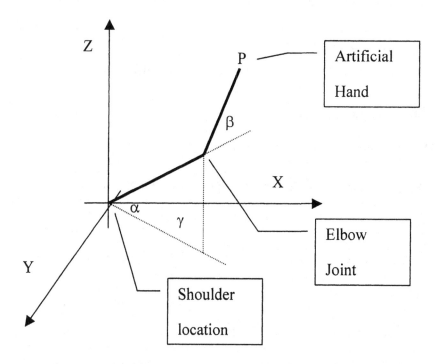

Figure 10.3. Scheme of a robot with an arm in a three-dimensional space.

For this the kinematics equation is:

$$\psi_1 = X = [\cos(\alpha) + \cos(\alpha + \beta)] \cos(\gamma),$$
$$\psi_2 = Y = [\cos(\alpha) + \cos(\alpha + \beta)] \sin(\gamma) \qquad (10.3)$$
$$\psi_3 = Z = [\sin(\alpha) + \sin(\alpha + \beta)]$$

Considering the two basis functions ψ_1 and ψ_2 where the angle γ is a variable of the robot we repeat the same process as the one used for the two-dimensional robot and we obtain the tensor g.

$$g = \begin{bmatrix} [\cos(\alpha)+\cos(\alpha+\beta)]^2 \int_0^{2\pi}\cos(\gamma)^2\,d\gamma & 0 \\ 0 & [\cos(\alpha)+\cos(\alpha+\beta)]^2 \int_0^{2\pi}\sin(\gamma)^2\,d\gamma \end{bmatrix}$$

$$= \begin{bmatrix} \pi[\cos(\alpha)+\cos(\alpha+\beta)]^2 & 0 \\ 0 & \pi[\cos(\alpha)+\cos(\alpha+\beta)]^2 \end{bmatrix}$$

The quadratic form is:

$$D(\alpha,\beta) = \pi\,[\cos(\alpha) + \cos(\alpha+\beta)]^2\,[\psi_1^2 + \psi_2^2]$$

So $[\psi_1^2 + \psi_2^2]$ which does not depend on the angle γ.

We have,

$$B2 = \psi12 + \psi22 = [\cos(\alpha) + \cos(\alpha+\beta)]\,2 \tag{10.4}$$

And,

$$D(\alpha,\beta) = \pi\,[\cos(\alpha) + \cos(\alpha+\beta)]^4$$

Which is invariant when the angle γ is changed. Using the new variable B the original system can be written as:

$$B = \cos(\alpha) + \cos(\alpha+\beta),$$

$$Z = \sin(\alpha) + \sin(\alpha+\beta) \tag{10.5}$$

This is similar to the equation system (10.1). So using the (10.5) and (10.3), we can calculate the angles α, β, γ when we know the variables X, Y, Z. In this way we solve the inverse kinematics problem for a three-dimensional robot.

The symmetry and second order action enable us to invert the non linear equation system (10.1) and the equation system (10.3). The inversion problem for a nonlinear equation system can be solved in an original way using the second order action and symmetry. For more information see John J. Craig [1].

10.2 Second Order Action and Population of Agents

10.2.1 Dynamical Programming

Given the states q_1, q_2,   ., q_n , we denote as π (q_i , n). This is the probability that the system will occupy the state q_i after n generations. We also denote $p_{i,j}$ as the probability that the state q_i will change into the state q_j . The next state function for calculating the probability of a particular state is given by the rule:

$$\pi(q_j, n+1) = \sum_{i=1}^{N} \pi(q_i, n) p_{i,j} \qquad (10.6)$$

The probability for the state q_j is equal to the sum of the products of the transition probabilities π (q_i , n) with the probability $p_{i,j}$ of a change of state from i to j . We have two probabilities: one is the probability for the state q_i at the step n, the other is the probability of moving from the state q_i to the state q_j Because we have many initial states q_i that can be transformed into q_j, we sum all of the possible contributions to obtain the probability π (q_j , n+1). For more details see [1]. In the expression 10.6 we must know the transition probability that is fixed for all of the processes. With transition probability we can describe a natural or industrial probability process when we know π (q_i , 0) and the transition probability. In this system where no input exists when the process begins we cannot control the system by using an input. A very simple example is the toymaker problem. The toymaker is involved in the novelty toy manufacturing business. He/she may be in either of two states. He is in the first state if the toy that he is currently producing has found great favor with the public. He is in the second state if his toy is out of favor. Let us suppose that when he is in state 1 there is 50 per cent chance of his remaining in state 1 at the end of the following week and, consequently, a 50 per cent chance of an unfortunate transition to state 2. Let us also suppose that when he is in state 2 there is 50 per cent chance of his remaining in state 2 at the end of the following week and, consequently, a 50 per cent chance of a fortunate transition to state 1.

Let,

$$p_{11} = 0.5 \;,\; p_{12} = 0.5 \;,\; p_{21} = 0.5 \;,\; p_{22} = 0.5 \;.$$

In a classical system the action of the input state can change into one and only one state. For a population of states one group can change in one way and another group can change in another way. So at the same time with different probabilities we can move from one state to two or more other states. In the population of the toys, it is possible to have two different states in the same environment. This is different from the case where we *have only one toy*. For one toy the only possibility is to move from state 1 to state 2 or in the opposite direction.

It is noted that when the proposition is:

P = " the toy is in state 1 ,

then the negation of the previous proposition is $\neg$ P = " the toy is in state 2 , so we have:

P and $\neg$ P

is always false. However when we have many agents or toys, the proposition P becomes:

P = " the toy X is in state 1

and at the same time the proposition

Q = " the toy Y is in state 2 ,

because Q is not the negation of the proposition P in the population we have no conflict.

So the transition diagram is:

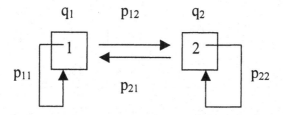

Figure 10.4. Transition of the states in the Markov chain or dynamical programming.

To describe the toy change of the states in the Toymaker Problem we include two different types of entities. In Figure 10.5 circles represent the first entity without any internal numbers or states. The circle entity represents anonymous persons who can change the state of the toys.

For the seller, in this particular case, it is not important to distinguish one buyer from the other. The second entity is represented by a square with two possible values which can be either one or zero. Number 1 if the toy has found favor, number 2 if the toy is out of favor. In Figure 10.5 we show 16 toys and 10 anonymous persons for the initial condition when all the toys are in the state 2. That is all the toys are out of favor.

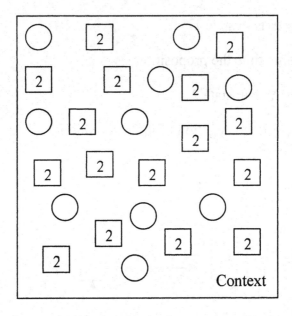

Figure 10.5. The Toymaker problem where we have 10 persons and 16 toys in state 2. All the toys are out of favor. This is the situation at time t = 0.

The persons can interact with the toys and can change the state with a given probability. In Figure 10.6 we show how the interaction with the persons changes from state 2 into state 1.

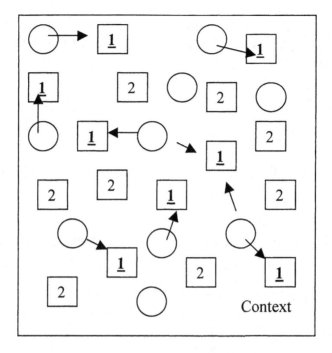

Figure 10.6. Showing the situation after one week: 8 toys (square) for the action of the customer (circle) change from state 2 (lack of favor) to state 1 (favored condition). The other 8 toys remain in state 2 after one week.

10.2.2 Evolutionary Computing

The simple genetic dynamical system is defined by A.E. Eiben: [3]. Here the system under consideration is defined in the following manner:

$$G = (\, SG\,,\, IG\,,\, OG\,,\, NG\,,\, RG\,)$$

$$SG = \{(N_1, N_2, \ldots \ldots, N_p\,)\}$$

$$\text{or } SG = \{(p_1, p_2, \ldots \ldots, p_q\,)\}$$

In the space of states SG is a q dimensional space of the number of copies N_k, also known as the population space. With a population

N_k we can normalize the population by substituting the probability p_k obtained by N_k divided by the total number of the elements in the population.

For the input space IG we have:

$$IG = \{(\alpha_1, \alpha_2, \ldots\ldots\ldots, \alpha_p)\}$$

Here $F_j = \alpha_j$ is the Fitness Function. The output is given as OG

$$OG = \{(ON_1, ON_2, \ldots\ldots\ldots, ON_p)\}$$

OG is the vector of the actions of every element or agent in the population N_k.

The next state function is given by:

$$NG = \{[(N_1, N_2, \ldots, N_p), (\alpha_1, \alpha_2, \ldots\ldots\ldots, \alpha_p)],$$
$$\beta (\alpha_1 N_1, \alpha_2 N_2, \ldots\ldots\ldots, \alpha_p N_p)\}$$

Where $\beta = \dfrac{1}{\sum\limits_{k} \alpha_k N_k}$ and α_k = fitness coefficient for the population

N_k, or,

$$NG = \{[(p_1, p_2, \ldots, p_p), (\alpha_1, \alpha_2, \ldots\ldots\ldots, \alpha_p)],$$
$$\beta (\alpha_1 p_1, \alpha_2 p_2, \ldots\ldots\ldots, \alpha_p p_p)\}$$

Where $\beta = \dfrac{1}{\sum\limits_{k} \alpha_k p_k}$.

Given the state of the population we define a function (Readout function RG) that from the state:

(N_1, N_2,........., N_p) gives the output (ON_1, ON_2,........., ON_p)

RG {(N_1, N_2,........., N_p) = (ON_1, ON_2,........., ON_p) }

In Figure.10.7 we show the agent interpretation of the fitness function which has four codes 00, 01, 10, 11 and four marks 1, 2, 3, 4 associated with every code. Each agent is represented by a square with two compartments. In the compartment on the left we show the code and in the compartment on the right we show the mark. In time the agents die at different rates. Agents with low mark die at high velocity, and agents with high mark die at low velocity. At every step both the probability and the number of agents are proportional to the mark of the agents. For example in the animal population, we observe that animals without illness or physical defects die at a low rate while animals with illness or physical defects die at a high rate. The physical performance is the mark of the agent animal. Agents are unable to change into other agents because there is not any reproduction. Moreover no form of communication exists between the independent agents to change the code or the mark.

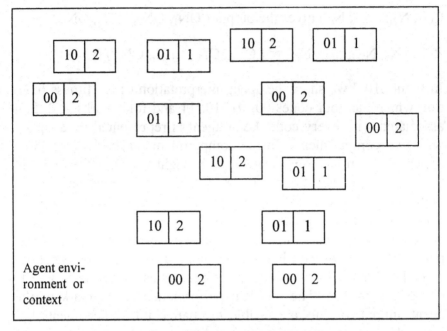

Figure 10.7. Every agent is represented by a square with two compartments. In one of the compartments there is a code, in the other a value (mark). The agent with low value dies at high velocity. The agent with high mark dies at low velocity. These codes cannot be transformed into other codes. In every generation we remove a part of agents with a low mark. After many cycles only the agents with high mark remain.

Following the description of a simple genetic dynamical system, we introduce the mutation operator within the genetic dynamics. Mutation is a process with random changes of the external agents.

Here μ is the probability to change and $(1 - \mu)$ is the probability not to change. For the four possible codes or chromosomes 00, 10, 01, 11 we have in Table 10.2 all the possible mutations. In Table 10.2 "No" means that no mutation has occurred while "Yes" means that a mutation of one part of the code has occured.

Table 10.2. Table of mutations for different codes.

Codes	00	10	01	11
00	No , No	Yes , No	No ,Yes	Yes,Yes
10	Yes,No	No , No	Yes,Yes	No ,Yes
01	No ,Yes	Yes,Yes	No , No	Yes , No
11	Yes,Yes	No ,Yes	Yes , No	No , No

For the mutation we have the table of probabilities:

$$U = \begin{bmatrix} (1-\mu)^2 & \mu(1-\mu) & \mu(1-\mu) & \mu^2 \\ \mu(1-\mu) & (1-\mu)^2 & \mu^2 & \mu(1-\mu) \\ \mu(1-\mu) & \mu^2 & (1-\mu)^2 & \mu(1-\mu) \\ \mu^2 & \mu(1-\mu) & \mu(1-\mu) & (1-\mu)^2 \end{bmatrix} \tag{10.7}$$

When the weights α_i of the fitness function are written as a diagonal matrix:

$$\alpha = \begin{bmatrix} \alpha_1 & 0 & 0 & .. & 0 \\ 0 & \alpha_2 & 0 & .. & 0 \\ .. & .. & .. & .. & .. \\ 0 & 0 & 0 & .. & 0 \\ 0 & 0 & 0 & .. & \alpha_p \end{bmatrix} \tag{10.8}$$

The new weights are $U\alpha$. The transition function of the genetic system is:

$$NG = \{[(\ p_1,\ p_2,\ ...,p_p\ ),(\ \alpha_1,\alpha_2,..........,\alpha_p\ )],\ \beta\ U\ \alpha\ p\ \}$$

Where,

$$\beta = \frac{1}{\displaystyle\sum_k \alpha_k p_k}.$$

For the matrices (10.7) and (10,8) we have:

$$U\alpha = \begin{bmatrix} \alpha_1(1-\mu)^2 & \alpha_2\mu(1-\mu) & \alpha_3\mu(1-\mu) & \alpha_4\mu^2 \\ \alpha_1\mu(1-\mu) & \alpha_2(1-\mu)^2 & \alpha_3\mu^2 & \alpha_4\mu(1-\mu) \\ \alpha_1\mu(1-\mu) & \alpha_2\mu^2 & \alpha_3(1-\mu)^2 & \alpha_4\mu(1-\mu) \\ \alpha_1\mu^2 & \alpha_2\mu(1-\mu) & \alpha_3\mu(1-\mu) & \alpha_4(1-\mu)^2 \end{bmatrix}$$

when $\alpha_1 = 3$, $\alpha_2 = 2$, $\alpha_3 = 2$, $\alpha_4 = 4$, the last line has the maximum sum or in other words the maximum probability. In this way the code 11 has the maximum of fitness and is the code with the highest probability of remaining after a number of generations. Mutation is a probabilistic input which changes the probability distribution. In Figure 10.8 circles are anonymous or unstructured agents as radiation or any agent that can generate mutation in the code. Squares are structured or principal agents with codes and marks.

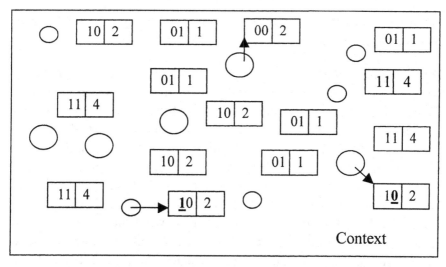

Figure 10.8. The Anonymous Agents (circles) communicate in a random way with the principal agents (squares) and change the code by using a mutation process. The code 00 appears as a mutation from the code 11. No meaning is contained within the communication between the Anonymous Agents and the principal agents with code and mark. The communication is a pure random communication without meaning.

When mutation occurs we move from one context to another. We show in Figure 10.9 the change of context.

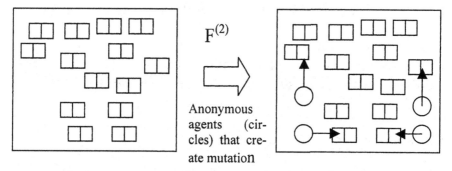

Figure 10.9. Second order action that changes the rules from a context without anonymous agents (circles) into a context with anonymous agents.

When two parents generate a child, the genes are mixed in a way that we describe as a crossover.

When we have the population codes 00,01,10,11 we can generate a new code or DNA from two codes or from DNA. Given the code 00 for the first agent and the code 11 for the second agent, the first part of the code for the first agent is 0 and the first part of the code for the second agent is 1. The probability that from the union of the two parts the child has the first part of the code equal to 1 is 0.5 because we have two possible cases 1 and 0 but we want only the value 1. For the second part of the code we have 0 for the first agent and 1 for the second agent and if we want the child to have 0 in the second code the probability is again 0.5 because we have again 0 and 1 and we want 0. In conclusion the probability to have the transformation $(00 , 11) \rightarrow 10$ is equal to:

$0.5 \ 0.5 = 0.25$.

this probability is denoted $\omega (e_i , e_j)_k$ where e_i is the number associated with the mother 's code, e_j is the number associated with the father's code, and k is the number associated with the child's code . For example when:

$00 \rightarrow 1 , 10 \rightarrow 2 , 01 \rightarrow 3 , 11 \rightarrow 4$, we have $\omega (1 , 4)_2 = 0.25$

where 1 is the number associated with the mother's code 00, 4 is the number associated with the father's code 11. 2 is the number associated with the child's code 10. We have the distribution of probability

$00 \rightarrow p_1 , 10 \rightarrow p_2 , 01 \rightarrow p_3 , 11 \rightarrow p_4$,

where $p_i = N_i / N$. Here N is the total number of the elements of the population and we have the vector C_k where:

$$C_k = \sum_{i,j} p_i p_j \omega(e_i, e_j)_k$$

The meaning of the C_k is that with a given child k, the probability is computed as the product of three parts. One is the probability for the crossing, the second is the probability of having a parent j and the third is the probability of having a parent i. We sum all possible couples of parents i and j.

With crossover and mutation from one code or DNA we can generate others codes or DNA.

The new transition function is:

$$NG = \{[(p_1, p_2, \quad ..., p_p), (\alpha_1, \alpha_2, ..., \alpha_p)], \beta^2 \ CU \ \alpha \ p \}$$

where C is the vector C_k.

In Figure 10.10 we show that two agents of low mark can communicate with a crossover in such a way as to generate agents with a high mark. The crossover is not a pure random process involving communication without meaning. The communication has a meaning in which a population with a high number of agents has a good chance to generate agents with a high mark. This also applies if these agents have a low mark. The crossover simulates a research in the space of all possible codes in such a way as to obtain agents with high marks.

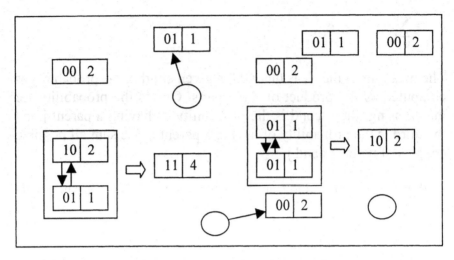

Figure 10.10. Here the agents communicate in a complex way to generate new agents that before were not present in the population. This type of communication is the crossover where one agent shares part of its code with another agent. Because the crossover is a process within the population, it is more sensitive to the probability distribution of the agents in the population with their codes. The communication between agents is here a structured communication with meaning. High numbers of agents but with low mark have a good chance to generate agents with high marks The circles are anonymous agents that generate mutations.

Figure 10.11 shows the second order action which proceeds to mutation and crossover from a simple genetic system.

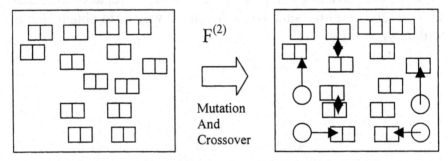

Figure 10.11. Second order action which introduces new rules into the new context by mutation and crossover.

In Figure 10.12 we describe a second order action by which we change the type of crossover. This action can be useful to adapt the crossover to different necessities. In many cases we can accelerate the crossover to facilitate the production of agent with high ability (mark) to solve special tasks. Second order actions can steer the crossover to adapt the agent population for different tasks to survive in difficult environments.

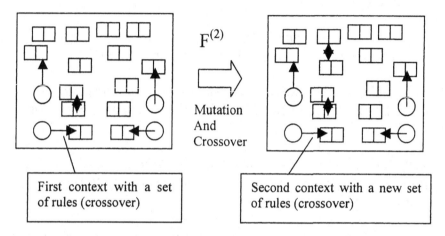

Figure 10.12. Second order action that changes the crossover rules.

In Chapters 2, 3 and 4 we have studied the action of the orders two, three and so on. When we use these actions to generate new crossover rules with possible internal incoherence, we extend the traditional genetic process from order one to a higher genetic order.

Bibliography

1. John J. Craig, *Introduction to Robotics*, Addison – Wesley, 1955.

2. Ronald A. Howard, *Dynamical Programming and Markov Processes*, MIT press, 1972.

3. Eiben A.E., *Theoretical Aspects of Evolutionary Computing*, Springer, Berlin, Heidelberg, New York, pp.13, 2001.

4. Jain, L.C. and Jain, R.K. (Editors), *Hybrid Intelligent Engineering Systems*, World Scientific Publishing Company, Singapore, 1997.

5. Jain, L.C. (Editor), *Soft Computing Techniques in Knowledge-Based Intelligent Engineering Systems*, Springer-Verlag, Germany, 1997.

6. Jain, L.C. and Fukuda, T (Editors), *Soft Computing for Intelligent Robotic Systems*, Springer-Verlag, Germany, 1998.

7. Jain, L.C. and Martin, N.M. (Editors), *Fusion of Neural Networks, Fuzzy Logic and Evolutionary Computing and their Applications*, CRC Press USA, 1999.

8. Lazzerini, B., D. Dumitrescu, Jain, L.C., and Dumitrescu, A., *Evolutionary Computing and Applications*, CRC Press USA, 2000.

9. Van Rooij, A., Jain, L.C., and Johnson, R.P., *Neural Network Training Using Genetic Algorithms*, World Scientific Publishing Company, Singapore, December 1996.

10. Vonk, E., Jain, L.C., and Johnson, R.P., *Automatic Generation of Neural Networks Architecture Using Evolutionary Computing*, World Scientific Publishing Company, Singapore, 1997.

Index